Change, especially cultural chang

change we must. And in changing, we must embrace the demographic changes that are happening in our world and on our campuses. *Leading a Diversity Cultural Shift in Higher Education*, with its rich case studies and analyses, provides a roadmap for building a campus where everyone belongs.

Robert S. Nelsen, President, California State University, Sacramento

It was a pleasure to have the opportunity to read *Leading a Diversity Cultural Shift in Higher Education*, by authors Edna Chun and Alvin Evans. Chun and Evans have provided a roadmap for institutions of higher education to transform the mindset of the academy to become more inclusive by inextricably linking the transgressions of our historical founding to the persistent challenges to achieving racial diversity and inclusion in modern day higher education. Let the robust dialogue now begin with a historical context in place!

Dennis A. Mitchell, DDS, MPH, Vice Provost for Faculty Diversity and Inclusion, Columbia University

Leading a Diversity Culture Shift in Higher Education is a must-read for any higher education administration practitioner or scholar. The book masterfully weaves illuminating anecdotes with the latest state of the art scholarship to provide a primer on the major diversity-related issues facing higher education leadership today. It provides a blueprint for thinking about positive institutional change.

Robert M. Sellers, Ph.D., Vice Provost for Equity and Inclusion & Chief Diversity Officer, University of Michigan

This research-based book by leading educators Edna Chun and Alvin Evans arrived at a critical moment in history, both domestic and international. A moment when we are experiencing a diversity cultural shift in higher education around the globe. Given my years of experience in the field of higher education and collaborative work with industry and healthcare organizations, this book is a must-read for practitioners, senior university administrators and industry executives.

Henry Odi, Deputy Vice President for Equity and Community and Associate Provost for Academic Diversity, Lehigh University

Leading a Diversity Culture Shift in Higher Education

Leading a Diversity Culture Shift in Higher Education offers a practical and timely guide for launching, implementing, and institutionalizing diversity organizational learning. The authors draw from extensive interviews with chief diversity officers and college and university leaders to reveal the prevailing models and best practices for strengthening diversity practices within the higher education community today. They complement this original research with an analysis of key contextual factors that shape the organizational learning process including administrative leadership, institutional mission and goals, historical legacy, geographic location, and campus structures and politics.

Given the substantive challenge of engendering a cultural shift for diversity in a university setting, this book will serve as a concrete primer for institutions seeking to develop a systematic and progressive approach to diversity organizational learning. Readers will be able to engage with provocative case studies that grapple with the current pressures emanating from diversity training and learn effective strategies for creating more inclusive environments.

This book is a perfect resource for institutional leaders, administrators, faculty members, and key campus constituencies who are seeking transformational change, institutional success, and stability in a rapidly diversifying national and global environment.

Edna Chun and **Alvin Evans** are award-winning authors and human resource and diversity leaders with extensive experience in complex, multi-campus systems of higher education. Two of their books, *Are the Walls Really Down? Behavioral and Organizational Barriers to Faculty and Staff Diversity* (2007) and *Bridging the Diversity Divide: Globalization and Reciprocal Empowerment in Higher Education* (2009), were the recipients of the prestigious Kathryn G. Hanson Publication Award by the national College and University Professional Association for Human Resources. They are also the authors of the first book appearing in Routledge's Critical Viewpoints book series, *Diverse Administrators in Peril* (2012). Their co-authored book, *The New Talent Acquisition Frontier: Integrating HR and Diversity Strategy in the Private and Public Sectors and Higher Education* (Stylus, 2014), received a silver medal in the 2014 Axiom Business Book Awards and is the first book to provide a concrete road map to the integration of HR and diversity strategy.

Chun and Evans are also regular contributors to a number of leading journals on talent management, HR, and diversity strategies. The co-authors are frequent presenters at national conferences and symposia including the College and University Professional Association for Human Resources, the Academic Chairpersons Conference, the Society for Human Resource Management, and the National Conference on Race and Ethnicity. Edna Chun is Chief Learning Officer and Alvin Evans serves as Higher Education Practice Leader for HigherEd Talent, a national human resources and diversity consulting firm.

New Critical Viewpoints on Society Series
Edited by Joe R. Feagin

Leading a Diversity Culture Shift in Higher Education
Comprehensive Organizational Learning Strategies
Edna Chun and Alvin Evans (2018)

Racism in the Neoliberal Era
A Meta History of Elite White Power
Randolph Hohle (2017)

Violence Against Black Bodies
Edited by Sandra Weissinger, Elwood Watson and Dwayne A. Mack (2016)

Exploring White Privilege
Robert P. Amico (2016)

Redskins?
Sport Mascots, Indian Nations and White Racism
James V. Fenelon (2016)

Racial Theories in Social Science
A Systemic Racism Critique
Sean Elias and Joe R. Feagin (2016)

Raising Mixed Race
Multiracial Asian Children in a Post-Racial World
Sharon H. Chang (2015)

Antiracist Teaching
Robert P. Amico (2014)

What Don't Kill Us Makes Us Stronger
African American Women and Suicide
Kamesha Spates (2014)

Leading a Diversity Culture Shift in Higher Education

Comprehensive Organizational Learning Strategies

Edna Chun and Alvin Evans

NEW YORK AND LONDON

First published 2018
by Routledge
711 Third Avenue, New York, NY 10017

and by Routledge
2 Park Square, Milton Park, Abingdon, Oxon OX14 4RN

Routledge is an imprint of the Taylor & Francis Group, an informa business

Library of Congress Cataloging-in-Publication Data
A catalog record for this book has been requested

ISBN: 978-1-138-28069-4 (hbk)
ISBN: 978-1-138-28071-7 (pbk)
ISBN: 978-1-315-21036-0 (ebk)

Typeset in Adobe Caslon Pro
by Apex CoVantage, LLC

To Alexander David Chun
"Alex"
Shining Star of Nobility

"When these elements come together, those of heart, soul and mind, individuals persevere and inner beauty is defined."

Alex Chun, 2002

CONTENTS

TABLES

FOREWORD

U.S. colleges and universities are usually seen as bastions of enlightenment and freedom, yet the many historically white institutions among them have long been central to the development of this society's systemic racism and other systems of societal oppression. This is a basic research-grounded premise of *Leading a Diversity Culture Shift in Higher Education*, a pathbreaking book by leading educators Edna Chun and Alvin Evans. In this savvy and very practical book they seek, successfully, to provide a framework for major changes in the institutional mindsets and unjust campus ecosystems of these historically dominant institutions.

Let us briefly reflect on some research buttressing this important premise, research that underscores how past societal conditions connect to contemporary issues of racial diversity and inclusion in higher education. Much of this research is largely unknown to most Americans. Consider, for example, the centrality of enslaved black labor to the construction and development of numerous private and public U.S. colleges and universities. Many had close ties to the slavery-centered economic system that dominated this country for the first 60 percent of its history. The first four founded in the colonial era—Harvard, William and Mary, Yale, and New Jersey—depended heavily on funding and other support from white slaveholders. As the historian Craig S. Wilder has made clear in his book, *Ebony and Ivy*, enslaved workers constructed campus buildings and provided student services. White trustees, administrators, faculty members, and students were often from slaveholding families. Over the next decades these private colleges were joined by numerous other private and public colleges and universities, in the North and the South—many also substantially dependent directly or indirectly on that slavery-based political-economic system.

This 246-year slavery era was followed by 90 years of Jim Crow segregation, eras together making up about 82 percent of this country's history. Then white Americans had exclusive, or much greater, access to first-rate educational programs as compared with Americans of color. Until the 1960s, most U.S. colleges and universities—except historically black colleges—were all-white or nearly so. Almost all historically white southern colleges and universities were racially segregated for this Jim Crow long era, while for much of this period most historically white northern colleges and universities also excluded, or had discriminatory quotas limiting, black students and other students of color.

Once in place, major aspects of the white-segregated institutions demonstrated great persistence over time, reflecting what might be termed the "law of social inertia." That is, in the social realm we observe a strong tendency for social oppression's exploitative mechanisms, resource inequalities, and buttressing framing to remain entirely or substantially in force until a major unbalancing force challenges that oppression. In the case of systemic racism, powerful whites have generally labored hard to keep that system from changing much, now over centuries of time.

When the system of racism does finally change, the law of social inertia operates to keep that system as much like the past as is possible. A major reason is that over the centuries the unbalancing counter-forces to the racist system, such as the major 1960s civil rights movements, have not yet been strong enough to force a large-scale dismantling of that oppressive system. Indeed, when under great pressures for change, whites in power have routinely preferred to make *limited* or *temporary* modifications, rather than to substantially or completely replace the deep racial and other oppressive structures of the United States.

Unsurprisingly thus, for centuries, elite whites, especially white men, have dominated most U.S. colleges and universities, and particularly at the top of historically white college and university hierarchies. They and their acolytes have controlled much of how colleges and universities are structured and operated, from staffing to curricula and degree requirements. Unsurprisingly, thus, a recent survey of major private universities found token-to-modest racial diversity in most senior administrative categories. Moreover, a disproportionate percentage of the administrators of color there were in diversity and student affairs positions. These professionals of color often have to signal to powerful white administrators and trustees that they will conform to the dominant white norms and expectations.[1] In a previous book (*Diverse Administrators in Peril: The New Indentured Class in Higher Education*) based on interviews with black, Hispanic, and Asian American administrators at major public and private universities, Chun and Evans found that most faced serious discriminatory barriers, including tracking into a limited number of administrative positions and a lack of decision-making authority over the scope of their supposed administrative authority. In addition, they often paid a high price in terms of stress and health problems that resulted from various racial barriers in their college settings.

As Chun and Evans regularly underscore here, the "continued dominance of white, male, heterosexual perspectives in university and college administration has failed to foster a representative bureaucracy that is responsive to the needs of diverse students" (Chapter 5, p. 221). This failure also extends to the perspectives of many senior white male faculty as well. Indeed, one goal of serious diversity and inclusion efforts is to make the faculties of our colleges and universities more representative of the country's population generally. As of now, nearly eight in ten full-time faculty members in our colleges and universities are white, and a majority of those who are full-time in the professor rank are white men.

Then there is the continuing contemporary problem of a great many whites college students' racist framing and actions. Most people familiar with the substantial research on the high level of white-racist framing and actions among white college students and other whites on and around college campuses doubtless understand why there is an urgent need to take serious diversity and inclusion actions at U.S. colleges and universities. Consider these rather commonplace white student actions in recent years: At the University of Connecticut law school, white students had a Martin Luther King, Jr., "Bullets and Bubbly" party where they engaged in aggressive racist stereotyping. White students put on baggy clothes and fake gold teeth, and held machine guns. At South Carolina's Clemson University numerous white students mocked the King holiday with a "ghetto-fabulous" party and dressed up in blackface and fake teeth grills.[2]

Over the last decade or two, a great array of racist events—including various "ghetto" and blackface parties, bake sales, and skits—have taken place at many other colleges and universities. The list includes, just to name a few, Washington University, New College, University of Tennessee, University of Texas, Trinity College, Whitman College, Willamette College, Texas A&M University, Emory University, University of Mississippi, Stetson University, University of Chicago, Cornell University, Swarthmore College, Tufts University, Massachusetts Institute of Technology, Macalester College, Johns Hopkins University, Oberlin College, Dartmouth College, Syracuse University, Tarleton State University, University of Colorado, University of Arizona, University of Alabama, University of Illinois, University of Delaware, and several University of California campuses.

These actions are not limited to just a few white students. In several studies of the racial views of white students that I and my colleagues have done, we have found that whites often engage in racist commentaries and performances in a variety of social settings. In a major study that Chun and Evans cite herein, titled *Two-Faced Racism*, Leslie Houts Picca and I analyze diaries about racial events, discussions, and performances that were kept by several hundred white students at numerous colleges and universities. The students recorded a very large number, indeed thousands, of clearly racist events going on around them. Many were in backstage settings with white relatives or acquaintances, while others were in frontstage settings with a racially diverse group of people present. These 21st-century diaries demonstrate how commonplace seriously racist commentaries and actions are among better-educated whites. African Americans appeared in a substantial majority of the white students' diary accounts about racist commentaries and performances, while other oppressed racial groups—e.g., Asian, Middle Eastern, and Latino Americans—were also periodically targeted. Significantly, less than 2 percent of these thousands of accounts showed any white person engaging in assertive dissent to racist activities. For the most part, these white college students viewed the racist commentaries and actions they or other whites do as being something that is rather normal, or at least relatively harmless.

As Chun and Evans suggest, these accounts of commonplace racist commentaries by white college students demonstrate not only how illiberal much of young white America still is on matters of racism, but also how much our historically white college and universities' white top administrators and senior faculty members have failed to engage enough in the educational, diversity, and inclusion efforts necessary to create really just and inclusive campus environments. Indeed, throughout this book, Chun and Evans accent the urgency of the reinvigorated and systematic organizational learning about diversity and inclusion that remains essential to major racial reform of our still racist white institutions of higher education.

Social science research demonstrates that such significant educational efforts can make a difference in everyday lives of white students and students of color. A growing number of social science studies reveal the importance and impact of diverse racial experiences and equal status contacts with regard to whites' racial framing of themselves and racial "others." In one study researchers compared white college students who scored high on certain antiracist measures with those who did not score well. Significantly, the more antiracist white students reported more diversity experiences growing up and on their college campuses. They also revealed in the testing more empathy for people of color targeted by racism. These researchers suggest a practical implication touched on by Chun and Evans, that white antiracist *role models* are very important in assisting other whites in confronting and dealing with persisting patterns of white racism.[3] Such role models can substantially impact student learning outcomes in regard to both on- and off-campus racial issues. In practice, college administrators and faculty leaders committed to major changes in institutional racial climates might make extensive use of antiracist white students and staff in campus diversity and inclusion efforts. This would be one important step in the direction of the essential and *systematic* organizational learning that is well assessed in this and other books that Chun and Evans have done on higher education.

Indeed, these social science research studies solidly back up basic conclusions that Chun and Evans emphasize in regard to needed college and university actions: First "that bold and courageous university and college leadership is required to overcome regressive internal and external forces resisting diversity change; and second, that systematic organizational learning is an indispensable lever for transmitting diversity across a campus landscape" (Chapter 1, p. XX). They thus insist there is no "quick fix" for the array of problems of diversity and inclusion on college campuses, that such changes require "sustained and deliberate institutional planning and action" (Chapter 1, p. XX).

Fortunately, since at least the 1960s, students of color and their supporters have played a key role in pressing this society's white leadership, inside and outside historically white educational institutions, to implement more organizational learning about diversity and inclusion and to institute some significant social justice changes in higher education. Current black and other progressive student movements on many college campuses such as Missouri, Yale, and Brown—including those triggered by the national Black Lives Matter movement—are underscored by Chun and Evans as having significant positive impacts in motivating some top white university administrators to take significant diversity and inclusion actions that improve their campus racial climates and programs.

Protests by black and other students of color have a long history in pressing for racial change at historically white educational institutions. For instance, one very important, and indeed continuing, impact of the 1960s black student movements was the creation of new African American courses in various academic disciplines and the generation of significant Black Studies programs at many colleges and universities. In his extensive research Ibram H. Rogers has shown that in that era hundreds of thousands of black and other student activists and their supporters at nearly a thousand historically white and black colleges and universities protested for African American cultural centers, academic studies programs, and changes in racist campus climates. The resulting organizational reforms brought more African American faculty and staff to campuses and changed, to varying degrees, traditional white-centered campus climates at historically white institutions. Nearly two hundred new African American studies programs facilitated the scholarly analysis of issues rarely analyzed and brought a black antiracist counter-framing of society to the center of many academic and public discussions.[4]

Undeniably, thanks in substantial part to these many and recurring student protests and demonstrations in the past and present, most historically white educational institutions now have racial-ethnic diversity and multiculturalism as, at least, their formal goals.

Nonetheless, in this systemically racist society there is an ongoing racial dialectic that inevitably includes much white pushback every time that significant progressive changes in racial norms or structures takes place. The contemporary white backlash against important academic innovations in this regard has resulted in much racialized retrogression. For instance, in recent years, African American studies faculty and students have often faced major cutbacks in these academic programs, as well as in related student support programs. The effect of these white-framed decisions, including those of powerful white college administrators and state legislators, is to move the country backwards on effective racial inclusion at historically white educational institutions.

In addition, since the 1970s, influential and rank-and-file whites have pressed for an end to many anti-discrimination and reparative programs (e.g., affirmative action) in various institutional arenas, including higher education. White students and their organizational backers have filed major law suits seeking to dismantle reparative programs that seek to redress centuries of white racial discrimination at historically white colleges and universities. For example, two cases involving the University of Michigan, both decided by a conservative-controlled Supreme Court in the 2000s, effectively set historical precedents limiting serious diversity and inclusion reforms in higher education across the country. More recent cases decided by the Court's conservative majority have continued to limit even modest use of race-conscious criteria in college admissions and other areas that try to address the long-term effects of centuries of systemic racism in higher education.

A major scholarly and practical contribution that Chun and Evans make herein is to insightfully review and fairly assess important case studies of particular universities where the racial dialectic is obvious, and especially where white-student and other white pushback against pressures from people of color for racial justice has increased. As noted previously, racial justice efforts by courageous student activists of color have long been major forces for progressive racial change inside and outside of historically white colleges and universities. However, the often powerful and negative white pushback—which, in the case studies in this book, demonstrably affects the ways in which many white administrators at historically white institutions respond—includes not only the resistance of powerful white alumni and other private white stakeholders, but also conservative white legislators, all of whom are explicitly or implicitly seeking to uphold a white-privileged status quo. Chun and Evans note that the hostility to increased campus diversity and inclusion efforts expressed by white political and civic leaders at the national level also makes such progressive change more difficult.

Implicit in much of the discussion and analysis in this important book is the pervasive white fear of ongoing demographic changes in the racial composition of the country's educational, economic, and political institutions. Many whites fear a much more multiracial future where they will be a statistical minority of the U.S. population. We observe these fears among both elite and ordinary white Americans. Prominent white commentators and academics have expressed very negative views about non-European immigration, multiculturalism, and the "browning of America" in leading publications and media outlets. For example, prominent Harvard professor, Samuel Huntington, argued in the influential journal, *Foreign Affairs*, that if multiculturalism becomes central to the United States, the U.S. could join other weak countries "on the ash heap of history."[5] The nativistic Huntington was explicitly concerned that contemporary immigrants are people of color drawn "overwhelmingly from Latin America and Asia." Although he offered several white nationalist arguments against this immigration, he did not deal in his analyses with the large-scale discrimination that non-European immigrants had received at the hands of white Americans for centuries.

These fears of non-European immigrants are seen among ordinary whites as well. Over recent decades, millions of whites have acted aggressively on their fears of racial change. Millions have moved from large cities with growing populations of color to whiter suburban areas, exurban areas, and gated central city communities. Most U.S. residential areas are, to a substantial degree, still racially segregated; and most white Americans and most Americans of color still live substantially separate lives in their local neighborhoods and schools.

Moreover, during the 2016 U.S. presidential election, the nativistic white developer Donald Trump routinely played on these white racial fears, including in regard to immigrants of color and people of color living in central cities. He regularly expressed a very negative racial framing of Mexican and Muslim immigrants and of black urbanites, and that racist framing helped him to win the presidency with a substantial majority of white voters. National surveys found that most of Trump's white supporters, often heavy consumers of nativistic or white nationalist mass media, were very disturbed by the country's growing racial diversity and racial-inclusion changes.

Significantly, one recent psychological study found that when white college students were shown U.S. population projections where whites are a minority, they became more angry and fearful of Americans of color, and even more white-oriented, than those not shown the projections in the study settings.[6] Just thinking about these demographic changes generated significant anger and fear among well-educated, mostly younger whites. Other studies have shown that whites of various ages and classes think, inaccurately, that whites are already losing much power and privilege in U.S. society to Americans of color and even that whites are now the victims of more racial bias and discrimination than Americans of color. Clearly, a great many envisage contemporary and future demographic changes from a white-racist framing of society that accents white superiority and virtuousness, and they do not value a

United States that will be much more diverse, minority white, and more just and equitable.

Nonetheless, these demographic changes are well underway, and more are coming. Today, white Americans are about 62 percent of the U.S. population, down significantly since 2000. Whites are currently a statistical minority in California, Texas, New Mexico, and Hawaii—and soon will be in numerous other states. For some years now, a majority of babies born in the United States have been children of color. U.S. Census Bureau estimates indicate that, assuming likely future birth and immigration rates, half the population will be Americans of color by the early 2040s, and perhaps before. By the 2030s more than half the U.S. working-age population will be people of color, and the composition of the U.S. voting population will likely be significantly different than today. In 1980 just 12.4 percent of voters in the U.S. presidential election were not white, but by 2012 that figure was 26.3 percent. By 2016 it had increased to 26.7 percent. In all these elections voters of color gave a substantial majority of their votes to Democratic Party candidates, including to the only U.S. president of color (Barack Obama) in both 2008 and 2012.

In future decades this demographic shift will almost certainly bring increasing and well-organized pressures from Americans of color for major changes in generations-old patterns of white discrimination and racial inequality in political, educational, economic, and other societal institutions. For instance, as voting majorities change from majority white, there will likely be significant changes in numerous white-controlled and white-privileged features of our governmental system—from jury composition and the operation of the criminal justice system, to the composition and operation of government agencies and state and federal legislatures, to the composition and operation of our public colleges and universities. How soon Americans of color become much more powerful in these institutions undoubtedly depends on the strength of their political and social coalitions and on the extent of anti-democratic countering measures taken by whites to maintain their socio-racial power and privileges. Given that white resistance to racial justice has grown and been rewarded since the white conservative Republican resurgence in state and federal governments over the last few decades, major societal changes in a social justice direction will likely be very hard to secure. Nonetheless, this change must come if that United States society is to survive and develop as a truly democratic country.

Very likely, too, change in the direction of greater societal justice and equity will also depend on the emergence and centrality of much more aware and progressive leadership for white America, leadership that recognizes that the country's *democratic* future lies in truly equitable and multiracial coalitions and institutions and in the full implementation of the U.S. rhetoric of liberty and justice for all.

Today, white leaders, as well as the white rank-and-file, cannot claim they have not been warned by many progressive analysts and provided with realistic equity solutions—especially, as in this book, for historically white educational institutions that have an obligation to educate the U.S. population, especially the white population, effectively for a democratic multiracial future. Indeed, in this book the ever-insightful educators Edna Chun and Alvin Evans draw on their own extensive experience in, and substantial research evidence from, contemporary U.S. colleges and universities to provide many *practical* and *well-tested* strategies for those administrators and faculty who do seek to make major changes in the diversity and inclusion realities on our still-troubled college campuses. They offer evidence, as they underscore, for effective "New leadership models that build teamwork, value cultural differences, create trust-based rather than fear-based environments, and reflect diverse constituencies are foundational components needed for substantive organizational change to occur" (Chapter 5, p. 221).

Joe R. Feagin
Texas A&M University

Notes

1. Patton, S. (2013, June 9). At the ivies, it's still white at the top. *The Chronicle of Higher Education*. Retrieved May 11, 2017, from http://chronicle.com/article/At-the-Ivies-Its-Still-White/139643/
2. Richard King, C., and Leonard, D. J. (2007, October 5). The rise of the ghetto-fabulous party. *Colorlines Magazine*. Retrieved May 22, 2009, from www.diverseeducation.com/artman/publish/article_9687.shtml
3. Kordesh, K., Spanierman, L., and Neville, H. A. (2013). White university students' racial affect: Understanding the antiracist type. *Journal of Diversity in Higher Education*, 6, 47–48.

4. Rogers, I. H. (2012). *The black campus movement: Black students and the racial reconstitution of higher education, 1965–1972*. New York: Palgrave Macmillan.
5. Samuel P. Huntington (September/October 1997). The Erosion of American National Interests, *Foreign Affairs* 76(5), p. 28.
6. Robert Outten, H., Schmitt, M. T., Miller, D. A., and Garcia, A. L. (2012). Feeling threatened about the future: Whites' emotional reactions to anticipated ethnic demographic changes. *Personality and Social Psychology Bulletin*, 38, 14.

ACKNOWLEDGMENTS

This book is dedicated to Alexander David Chun who left us all too soon and whose beautiful, generous spirit, courage, and passion for music, the environment, social justice, and all living creatures are a living legacy of hope and truth.

We are deeply appreciative of the guidance of Professor Joe R. Feagin, Ella C. McFadden Professor of Sociology at Texas A&M University, throughout the writing of this book and his encouragement to pursue the topic of diversity organizational learning in higher education. We would like to express our sincere gratitude to Dean Birkenkamp, Senior Editor, Sociology, at Routledge for his vision and help in bringing this work to fruition.

We thank all the college and university leaders, diversity officers, faculty, administrators, and students who we interviewed for this study for their inspiring testimonials and unflagging dedication to the work of leading a diversity culture shift. While these contributions are too numerous to identify, we especially would like to mention the invaluable insights and assistance of Professor Charles Behling of the University of Michigan (Retired); Bryan Cook, Senior Vice President for Educational Research and Analysis at the American Dental Association; Michele Minter, Vice Provost for Equity and Diversity at Princeton University; Professor Ted Morgan of Lehigh University; Henry Odi, Deputy Vice President for Equity and Community and Associate

Provost for Academic Diversity at Lehigh University; and Professor Berkeley Hudson at the University of Missouri at Columbia. We also would like to express our appreciation to Kimberly Thompson Rosenfeld for her skilled research assistance.

Alvin Evans would like to thank his children, Shomari Evans, Jabari Evans, Kalil Evans, and Rashida VanLeer for their continuous and loving support. Edna Chun thanks Jay K. Chun, David S.C. Chu, and George S.T. Chu for their ongoing care and loving help.

1
AN IMPROBABLE LANDSCAPE FOR DIVERSITY CULTURE CHANGE

> On campuses today you see more bonding than bridging. If we do not deliberately create conditions to encourage real bridging, we are left with a series of disconnected conversations that do not feed or challenge one another, a kind of archipelago of islands of like individuals occupying the same ocean, and not a community forged of common ground.
>
> Julio Frenk, President of the University of Miami, 2016, para. 12[1]

In March 2007, then Senator Barack Obama who later became America's first biracial president, described the civil rights journey begun by four students from North Carolina A&T University. The African American student activists, dubbed the Greensboro Four, dared to sit in at a lunch counter in a Woolworth's in Greensboro, North Carolina:[2]

> The previous generation, the Moses generation, pointed the way. They took us 90 percent of the way there. We still got that 10 percent in order to cross over to the other side. So the question, I guess, that I have today is what's called of us in this Joshua generation.

In this biblical reference, Obama viewed the historic journey toward inclusion as the unfinished work of the Joshua generation. While Moses enabled the Hebrews to leave Egypt and nearly reached the Promised Land, Joshua had the task of helping the Hebrews cross over the Jordan River.[3]

Students representing the new Joshua generation have pressured for diversity progress in higher education. In fact, historians may view this time period as one in which young people drove the process of cultural change, a time when the dimension of youth intersected with race, class, and identity.[4] Higher education will undoubtedly play an "outsize role" in the historical record, as student demonstrations in higher education crystallized the "explosive strands" of demography and social change in the United States.[5]

Given the unfinished journey that former President Obama describes, how can colleges and universities foster more inclusive learning, living, and working environments for faculty, staff, administrators, and students? What contextual conditions and leadership strategies will lead to a comprehensive diversity culture shift and offer the greatest potential for success?

Scholars indicate that colleges are changing and can indeed change,[6] while skeptics see diversity progress as slow, even glacial, and marked with the type of resistance that accompanies the effort to preserve well-established patterns and practices. Few institutions have implemented successful approaches to diversity learning that transcend organizational silos and transmit diversity learning across the varied counters of campus topography. More commonly, diversity activities and programs are fragmented, piecemeal, and lack sustained and integrative programmatic impact.

Diversity change is not a "quick fix" but requires sustained and deliberate institutional planning and action. Committed diversity leadership needs to shift the tides of institutional culture and navigate the rocky course of changing institutional mindsets in order to embed the value of diversity throughout a campus ecosystem. As demonstrated by the interviews and case studies in this book, the progressive course of diversity organizational learning is subject to many variables and attainment can fluctuate based on institutional leadership and cultural readiness.

To provide a compass for the change process, this book offers a research-based framework for initiating, implementing, and sustaining a diversity culture shift. To date, only a small body of research addresses how systematic diversity organizational learning can be implemented in higher education. The purpose of this study, using the analogy suggested by Julio Frenk, is to find strategies that will bridge the archipelagos of islands that co-exist independently within a college or university, build community, and transmit shared diversity values throughout the institutional ecosystem. As Frenk explains, institutions of higher education need a "scholarship of belonging" that embraces the multiple dimensions of diversity and links "our *value* to our *values*."[7]

Our central premise is based on two fundamental assumptions: first, that bold and courageous university and college leadership is required to overcome regressive internal and external forces resisting diversity change; and second, that systematic organizational learning is an indispensable lever for transmitting diversity across a campus landscape. The dynamics of a diversity culture shift depend upon committed leadership, a governance infrastructure that supports change, and clear institutional direction.

In a larger sense, the higher education environment represents a microcosm of the structural, institutional, economic, political, cultural, and psychological conditions in society as a whole that lead to powerlessness.[8] Institutionalization has reproduced and replicated forms of exclusion and oppression of members of nondominant groups that constitute a longstanding "material, social, and ideological reality" in American society.[9] The five case studies of public and private re-search universities shared in this book reveal the ways in which institutional cultures can transmit, reproduce, and perpetuate the exclusionary norms, attitudes, and practices present in wider society. For this reason, we adopt the terminology of minoritized or nondominant groups to address groups that have wielded or held less power respectively within institutional settings based on salient demographic characteristics that include race, ethnicity, gender, sexual orientation, gender identity, and disability status as well as the intersectionality of these dimensions.[10]

During the interview process for this study, a question we often received from campus leaders was, "What do you mean by organizational learning?" As Nicole, a diversity dean at an elite university observes, at her institution faculty remain skeptical about knowledge the institution has not produced itself:

> There is nobody here who has a concept of organizational learning. . . . We are in a place of very, very smart people who I think care in some very abstract way about this issue . . . but are unconvinced that they don't already have the answers to the problem . . . we are functioning at a level that is not as strategic as it is reactive. I don't think we are unique.

The case studies in the book illuminate the ways in which the leadership structures of higher education have fared under seemingly antagonistic forces. These forces include the following factors: 1) increased intervention of conservative state legislators in the governance of public institutions of higher education, 2) hostility to diversity and divisiveness arising from national political events, 3) skepticism of alumni and external stakeholders about student demonstrations, and 4) pressures from minoritized student groups seeking to create more inclusive campus climates.[11] In the most extreme example at the University of Tennessee at Knoxville (UTK), legislative attacks have resulted in the withdrawal of diversity funding and led to administrative turnover. The case study at the close of Chapter 2 illustrates the dramatic tension between student demands and the legislators' view that UTK had deviated from "Tennessee values" or the values of the majority. At the same time, a clear trend that emerges from the troubled institutional waters of higher education is the voice of tenured faculty leadership that in some instances has garnered strength through faculty senates.

Our examination of the institutional landscape for diversity foregrounds the critical role of systematic organizational learning in diversity transformation. From this perspective, organizational learning needs to be a core element of an institution's diversity agenda, an agenda that integrates diversity and inclusion into the organizational structure and culture of the college or university.[12] This book addresses the specific ways in which organizational learning can serve as a conduit for changing mindsets, assumptions, and prevailing norms related to diversity within the fabric of institutional culture. Most importantly, we share approaches that lead to substantive institutional outcomes.

The Tumultuous Political Stage for Diversity

Beginning in late 2014 and continuing well through the fall and winter of 2015, a wave of student demonstrations swept college campuses around the nation. The protests initially arose following the killing of an unarmed African American teenager, Michael Brown, in Ferguson, Missouri by a white police officer in August 2014 and the determination in November of that year that no indictment would be issued to the officer.[13] A tsunami of student protests was ignited by demonstrations at the flagship University of Missouri campus at Columbia ("Mizzou"), just 116 miles away.[14] The Ferguson incident triggered discontent on the Mizzou campus regarding a perceived legacy of marginalization and exclusionary treatment of minoritized students. In the case study at the close of this chapter, we describe the aftermath of the student demonstrations at Mizzou and the subsequent administrative upheaval.

Concerns about the persistent lack of inclusion for members of nondominant groups on predominantly white campuses were at the forefront of student concerns. Student activism expressed impatience regarding the pace of diversity progress. In a seemingly domino effect, across the spectrum of college campuses including Ivy League institutions, research universities, and private liberal arts colleges, students demanded changes in the campus environment for diversity. These demands not only emphasized the need for greater diversity among faculty, administrators, staff, and students, but also called for leadership support, policy changes, and training programs that would create more welcoming living and learning environments.

A survey of 76 institutions and organizations reveals remarkable commonality in the themes that student demonstrations identified: 91 percent of student groups called for policy change; 89 percent had specific demands for campus leadership practices; 88 percent requested greater resources be devoted to the needs of diverse students; 86 percent sought greater structural diversity among students, faculty, staff, and administration; and 71 percent requested new or revised cultural competency or diversity training.[15] While the demands for training were varied, students emphasized the importance of systemic training on diversity and cultural competency for all campus constituencies including faculty, staff, students, administration, and police. Further, students asked institutions of higher education to adopt significant shifts in organizational culture instead of implementing stand-alone diversity initiatives.[16]

Cynics viewed student demonstrations around the nation as immature and as simply "another skirmish" in "the tedious culture war" that has gone on for decades.[17] Alumni of institutions such as Amherst College and Yale University described how they were baffled by the culture of higher education, feeling that students are engaged in identity politics and allowed to pursue frivolous curricular interests.[18] These alumni expressed their disapproval by withholding their donations such as Scott MacConnell, an alumni of Amherst, who voiced his displeasure as being "lied too, patronized and basically dismissed as an old, white bigot" who does not understand the needs of the current campus community.[19]

At some institutions, symbolic administrative gestures or "concessions" have failed to trigger substantial diversity change and seem comparable to simply treading water or running in place. At other institutions the impetus of student demonstrations has resulted in new budgetary allocations and a more systemic review of diversity policies, practices, and programs. As institutions of higher education analyze the current diversity landscape and develop new approaches, the pathway forward is often perilous, characterized by hidden minefields with few benchmarks or exemplars available. In a highly decentralized campus environment, even communicating a common vision for the change process across the decentralized institutional landscape is a daunting prospect.

Catherine, an African American diversity dean at an elite eastern liberal arts college, explains how student pressures have created greater momentum for change. Like other diversity officers we interviewed, she sees the need for continued and more rapid change:

> Compared to other small liberal colleges, I think we are in a good place, but we also know that we have to do better. We know the

> demands of history require that we do better for our students. We are pushed by our students as much as we are pushed by faculty and staff. . . . Sometimes the students push us to move faster than we are prepared to go, but we know we have to continue to move forward, perhaps with a little more swiftness or haste to keep up with the times in which we are living.

A number of well-funded public and private institutions have implemented sweeping new diversity plans with unprecedented budgetary commitments. In some cases, the rapid development of diversity planning documents and infusion of funds immediately followed student protests and demands such as at Brown University and Yale University. We share here three examples of major diversity initiatives implemented since 2015 at Brown, Yale, and the University of Michigan.

The Brown Response

On November 16, 2015, Concerned Graduate Students of Color at Brown University presented a list of unmet demands, requesting a response within one week.[20] Specific areas of concern noted by the students were the use of voluntary diversity training modules rather than mandatory inclusion and anti-oppression training and the need for tangible reporting systems that address hiring and recruiting practices and departmental climate. Just three days later, President Christina Paxson's Office produced a working draft of a Diversity and Inclusion Action Plan promising a $100 million investment in diversity.[21] The final version of the plan, "Pathways to Diversity and Inclusion (DIAP)" was launched in early 2016 with an announcement of a $165 million investment.

Brown's plan is closely allied with its strategic plan, "Building on Distinction," released in 2015 that identified $100 million for new endowed faculty positions. The stated goal of DIAP is to hire 25 percent of these positions from diverse groups and to use existing faculty lines to hire 60 faculty members from historically underrepresented groups over a five- to seven-year period.[22] Twenty-five million dollars will be devoted to diversifying the graduate student population.

It is important to note that Brown defines historically underrepresented groups (HUGs) as individuals who self-identify as American

Indian, Alaskan Native, African American, Hispanic or Latino/a, and Native Hawaiian and/or Pacific Islander. Asian Americans are only identified as underrepresented in the humanities and social sciences and women in STEM fields. This designation derives from a common misconception that Asian Americans are overrepresented throughout the academy often as a result of the model minority stereotype. In fact, Asian American women are underrepresented in almost every field.[23] Further, the statistics on Asian American faculty reflect a substantial international population and although Asian Americans represent 8 percent of the professoriate, an estimated 40 percent are foreign nationals based on data gathered by the Equal Employment Opportunity Commission from 1981 to 1991.[24] Asian American faculty, like other minoritized faculty, often face a glass ceiling, chilly departmental climates, and exclusionary behaviors. A study based on the 2004 National Study of Postsecondary Faculty (NSOPF) involving 26,100 faculty at two- and four-year institutions found that being Asian American Pacific Island (AAPI) was associated with a 56 percent decrease in the likelihood of attaining a tenured faculty position.[25]

Given the trend by a number of universities to exclude Asian Americans from the designation as a historically underrepresented group, Frank Wu, a noted scholar and lawyer, identifies the absence of data that can accompany such determinations. As he explains:

> The stereotype is that Asian Americans don't face stereotyping, so when Asian Americans complain, the response is, "Well we all know you are doing well." And it becomes a way to limit Asian American advancement, because Asian Americans actually are underrepresented: we are not in the room when these discussions occur, and can't advocate for ourselves, and it's is not in the interests of even people who might be our allies to advocate for us. They have no reason to, because this is sometimes perceived to be a zero sum game.[26]

A number of features of the Brown plan offer significant initiatives that advance diversity organizational learning. From a scholarly perspective, the plan identifies new research initiatives such as the allocation of $20 million to research centers focused on race, ethnicity, and social justice. The plan requires departments to submit annual diversity and inclusion plans with the expectation that departments will report on participation in diversity professional development. It also includes required training on diversity and inclusion for all members of tenure, promotion, and appointment committees (TPAC) and expansion of voluntary diversity training programs. In addition, the Office of Institutional Diversity and Inclusion (OIDI) will implement a Diversity and Inclusion Professional Development Collective (D&I Collective) that utilizes in-house faculty, administrators, staff, and students as trainers and consultants.

The first departmental diversity plans were submitted in 2016 following the rollout of the "Pathways to Diversity and Inclusion Plan" (DIAP). Liza Cariaga-Lo, Vice President for Academic Development, Diversity, and Inclusion, indicates that the central repository of departmental plans will serve as a living and working resource that will be revisited annually. The goals established in the departmental plans coupled with feedback from focus groups, faculty, staff, and students offer a substantial framework for future professional development planning. As Cariaga-Lo, an Asian American female, candidly explains:

> we are trying to do a lot of experimentation around what can work best and what doesn't, and we are trying to lead efforts that have promise and recognize that this will take some time for us to be able to fully reap the benefits of the kind of diversity and inclusion professional development initiatives that we are putting forward.
>
> It's a work in progress and we're actively assessing on a regular basis all that we are trying to do and putting forward a set of metrics that will allow us to measure the progress that we are making on a number of different fronts. I certainly am cautiously optimistic that this time around we are being much more intentional in fulfilling the goals that we stated and that are in our DIAP documents.

The Yale Response

A similar scenario unfolded at Yale University where two weeks of student protest in fall 2015 reflected growing racial tensions, culminating in a march of hundreds of student protesters on November 6, 2015. Just three days earlier, on November 3, 2016, President Peter Salovey announced that $50 million would be allocated over a five-year period to recruit exceptional faculty who enrich diversity or another area of strategic importance and would expand faculty development offerings in the area of diversity.[27]

Following the protests, in May 2016, the Senate of the Faculty of Arts and Sciences (FAS) released a rather critical report on faculty diversity and inclusivity that called on university leadership to "articulate a cogent vision for the intellectual value of diversity and inclusivity as well as a clear plan of action."[28] Notably, however, the report does not identify the definition of underrepresented minority (URM), but instead refers to the distinction between "federally mandated guidelines on populations that are underrepresented" versus minority faculty not recognized as URM's and international faculty. The reference to federally mandated guidelines is unclear, since federal affirmative action guidelines identify all minority groups in a single category.

The FAS report warns against "a disproportionate emphasis" on implicit/unconscious bias which could overshadow the need for attention to structural, leadership, and budgetary issues related to diversity. It indicates that an emphasis on implicit/unconscious bias could alienate underrepresented faculty by not addressing more systemic forms of exclusion.[29] This insight has significant ramifications for diversity organizational learning when workshops and seminars focus on implicit/unconscious bias without attention to the structural underpinnings that perpetuate forms of exclusion.

Structural initiatives implemented at Yale include the formation of the Presidential Task Force on Diversity and Inclusion with the president serving as chair and the creation of a deputy dean for diversity in the College of Arts and Science. Specific organizational learning programs include diversity training for the university's top administrators; leadership training on recognizing and combatting discrimination; a five-year series of conferences on race, gender, inequality, and inclusion; and training on diversity and inclusion in the freshman orientation program. A new center on the study of race, ethnicity, and social identity will also be established.[30]

The University of Michigan's Expansive Diversity Plan

One of the most ambitious and systematic diversity planning initiatives in public higher education was launched at the University of Michigan (UM) under the leadership of President Mark Schlissel, the first physician-scientist to lead the university, who was appointed in 2014. The diversity strategic planning process began shortly after Schlissel's appointment in early 2015, following the recommendations of a task force reporting to the provost which had been convened earlier. After a lengthy planning process and a week-long diversity summit, a five-year diversity strategic plan was launched officially in October 2016. The $85 million plan added $40 million to the funding already devoted to diversity. The fact that the university is constitutionally independent from the state legislature has enabled it to attain greater programmatic autonomy in diversity compared to public universities in some other states.

The UM plan centers on three strategic goals: 1) creation of an equitable and inclusive campus climate, 2) recruiting and retaining a diverse community, and 3) supporting inclusive and innovative scholarship and teaching.[31] As part of the expansion of the university's diversity program, Robert Sellers, a tenured professor of psychology and former vice provost for diversity, equity, and inclusion, was named as the university's first Chief Diversity Officer.[32]

The university's diversity plan is organized as a distributive model with 49 academic and administrative units responsible for developing their own plans. The first plans were submitted in the fall of 2016 in alignment with the university's three overarching diversity goals. Units held focus groups and town hall meetings to gather input for plan development. The plans were then reviewed by the units' respective organizational areas including academic affairs, business/administrative area, health centers, and athletics. Given the quick turnaround in the first year, Sellers describes the initial plans as "a bit uneven with respect to the ambitiousness of the goals articulated in them. But I think they all reflect the vision and goals of our community and have addressed the questions that they were asked to address."

According to Sellers, the goal of the UM Diversity Plan is proactive, not reactive and focused on long-term change. He also notes that UM's students "have been very active over the years and have stayed on the forefront" and underscores the impact of courageous administrators on diversity progress. Sellers further indicates that President Schlissel views the attainment of a more inclusive campus as a key measure of his success. The diversity plan has received strong support from the university's Board of Regents. As Sellers explains:

> The goal was to be proactive and responsive, but not reactive. And we wanted to have enough time to really think about what we were hoping to accomplish. Our focus was on long-term institutional change, recognizing that meant that the goal was not simply to reduce student unrest, and so success wouldn't be defined in terms of whether or not there were student protests.
>
> The president immediately embraced the recommendations of the task force and had a strong commitment to engaging in the diversity strategic planning process. From the start, he said that it was something his presidency should be judged on. He stated that if we had not achieved progress in regard to diversity, equity, and inclusion over the term of his presidency, it would be a failure. That made things so much easier. The other important factor was that the board of regents was also very committed.

The UM plan offers a multifaceted and differentiated focus to organizational learning for students, faculty, administrators, and staff. A diversity, equity, and inclusion education training resource has been developed which currently has eight programs designed to develop behaviors and skills that will help improve campus climate.[33] The university also named its first chief organizational learning officer housed within Human Resources and focused on building capabilities of staff, managers, and leaders throughout the university.[34] In addition, the university will expand its Inclusive Teaching Professional Development Program and convene working groups to determine how diversity, equity, and inclusion-related contributions can be incorporated into the tenure and promotion process and staff evaluations.[35] Professional development for new deans and executive leaders related to diversity and inclusion will be provided as part of the on-boarding process and also addressed during the executive and deans' retreats.

These three examples of comprehensive, university-wide diversity initiatives at Yale, Brown, and Michigan represent significant investments that offer promise for sustained change and enhanced diversity organizational learning programs.

Jim Crow Revisited

At the same time as demonstrations on college campuses have drawn attention to the persistence of exclusionary practices, deepening divisiveness over issues of race and ethnicity simultaneously has polarized the national landscape. This overt hostility can be understood as surfacing deeply buried racial antagonisms and anti-immigrant sentiments by white, working-class Americans to the surface.[36] In a supposedly post-racial era, deep-seated frustration and anger regarding the "ultimate affront," the election of Barack Obama, a biracial individual as president of the United States for two successive terms in 2008 and 2012, gave rise to white rage.[37] This rage derives from the pairing of blackness with advancement, ambition, aspirations, and demands for equal citizenship.[38] Republican commentators and officials repeatedly directed mock racial framing at Obama.[39] The racial divide resulted in "a formidable array of policy assaults and legal contortions" that, in the words of one critic, "consistently punished black resilience, black resolve."[40]

The nation's divisive political climate was further fueled by the racist, misogynistic, and xenophobic rants and rhetoric of Republican presidential candidate Donald J. Trump who subsequently was elected to the presidency in 2016. Trump's rhetoric appeared to move the clock backward to erode the hard-won gains of the Civil Rights movement. The intense racialized polemic generated by Trump suggested a return of the segregated period of Jim Crow (1880–1969) that occurred in southern and border states. With the anti-immigrant, sexist, and white racial framing of Trump's public messages, the landscape for diversity on college campuses became even more divisive and problematic.[41] The dramatic collision between the American political macrosystem with the dynamic microsystems of college student demonstrations that arose in 2014 and 2015 is only certain to intensify.

With the demographic decline of whites in America, Donald Trump has given voice to the anger of white citizens who do not feel privileged.[42] His rhetoric crossed the boundaries that have restrained America's discussion of race and unleashed the anxiety driven by a sense of white vulnerability.[43] Drawing on President Nixon's reference to the silent majority, Trump appealed to those Americans who sought the return of the silent majority, which many interpret as a reference to white citizens taking their country and culture back.[44] Trump's call for a temporary ban on entry of Muslims who are non-citizens, tepid disavowal of the Ku Klux Klan, and referral to Mexicans as rapists have been compared by NAACP president William Cornell Brooks to "a kind of Jim Crow with hairspray and a blue suit."[45] In an era in which racism, anti-immigration sentiments, and fear of Islam have increased, Trump's reactionary pronouncements downplay racist and nativist tendencies that demonize whole groups of people by reducing them to objects, giving such tendencies secular and ethical legitimacy.[46]

The name "Trump" has been used as a taunt associated with racist rhetoric on campus, such as when two students at Northwestern University were charged with spray painting a college chapel with slurs against African Americans and LGBT individuals with the word "Trump".[47] In another example, a Muslim and a Hispanic student at Wichita State University were verbally assaulted by a man at a gas station shouting, "Trump, Trump, Trump, we will make America great again. You losers will be thrown out of the wall".[48] In these cases, as social theorist Joe Feagin points out:[49]

> They are chanting like they do a football games for their "team," and their strong "leader"—in this case Trump has become their team leader and "coach," for "team -angry white" or "team angry white male".

The examples reveal how the invective of Donald Trump has "normalized" racism and xenophobia in day-to-day interactions both in the public space and on campuses.[50] In a survey of 2000 K-12 teachers, more than two-thirds of these educators reported the alarm that this campaign rhetoric has caused among children of immigrants, racial minorities, and Muslims.[51] Since one-third of American students have foreign-born parents, vulnerable students are fearful, stressed, and hurt not only by Trump's ideas and tone, but also by the number of people that seem to agree with him.[52]

On college campuses and in public spaces, the "Trump effect" has resulted in the movement of overtly racist behaviors and comments from backstage settings among white actors to the frontstage where diverse audiences are present.[53] In the past, diversity discussions on campus have often taken place in veiled and muted rhetoric that mask the troubling social realities of racism, sexism, heterosexism, gender discrimination, and ableism that thread American society. For example, as shown in a survey of 626 white students at 28 colleges and universities, in the frontstage, diversity is discussed in polite terms before mixed audiences, while in the backstage students recorded 7,500 instances of racist comments and events.[54] By March of the 2016–2017 academic year, the Anti-Defamation League reported 107 incidents of fliers distributed by white supremacists on college campuses. This uptick in white supremacist actions was attributed to these groups feeling emboldened by the national political climate.[55]

As a result, even despite the deluge of student protests regarding race relations on campuses across the country, the national mood in the country has not been supportive of diversity change. As commentators noted, "the upheaval morphed into a Rorschach test that reflects the complex state of American race relations."[56] Rather, diversity resistance has coalesced in the actions of conservative state legislatures that have used the threat of budgetary reduction to cross into the realm of public university governance through an explicit critique of appointments, programs, policies, and practices. Such political pressures have had a demonstrably chilling effect on the tenor of diversity programs, even when such programs were initiated by student groups. Regressive legislative action and divisive national rhetoric fly in the face of one of the most frequently cited arguments for diversity progress: the changing demographics of student enrollment.

The New Demographics: A Litmus Test for Change

Much has been written about the "diversity explosion" that is propelling the U.S. population toward a minority majority nation by 2044 in which no racial group will be the majority.[57] Two countervailing population trends are projected to continue from 2014 until 2060: a decline in the white population and the growth of new minority/ethnic populations including Asian and Latino/a populations, which will likely double, and multiracial populations that will likely triple.[58] Even so, demographic change often is presented in a neoliberal framework that emphasizes diverse markets and "the business case" for diversity and fails to address the impact of asymmetrical power structures within institutional contexts that privilege white, male, younger, non-disabled, heterosexual males.[59] As we shall see in the case studies in this book, the implications of the demographic shift for diversity research necessarily involve the relationship between diversity and the twin factors of power and context.[60] Given the transient nature of student populations, student demonstrations inevitably will lack sustainability without leadership direction from the campus power structure.

Ignoring the shift in student demographics is no longer an option for institutions of higher education. The number of racial and ethnic minority students is projected to rise significantly between 2010 and 2021 with a growth rate of 25 percent for African American students, 42 percent for Hispanic Students, 20 percent for Asian/Pacific Islander students, and only 4 percent for white students.[61] Table 1.1 below reveals

Table 1.1 Changes in Race/Ethnicity in Undergraduate Student Enrollment

Race/Ethnicity	*2004*	*2014*
White	63.9%	55.6%
Asian	5.7%	6%
African American	11.1%	11.8%
Hispanic	9.2%	13.5%
Two or more races		3.1%
Native Hawaiian/PI		0.3%
Non-resident Alien	2.7%	4.3%
Unknown	6.4%	4.7%

U.S. Department of Education, 2004, 2014, Washington, D.C.: National Center for Education Statistics, Integrated Postsecondary Education System. Analysis by authors.

the impact of demographic population shifts with a significant increase in racial/ethnic minority undergraduates enrolled in four-year institutions, master's, and doctoral level universities and an 8.3 percent decline in the number of white students over a decade.[62]

The complexity of the college and university enrollment picture also includes growth in non-traditional, first-generation, and socioeconomically disadvantaged students. First-generation college students are more likely to be Latino/a or African American, female, and financially self-reliant. Many first-generation students enroll in college part-time, live off campus or at home, and attend two-year institutions.[63] Further, demographic changes also necessarily include gay/lesbian/bisexual and transgendered (LGBT) students. Since LGBT students often face the dilemma of whether or not to disclose their identity, statistics on these individuals are not readily available. Nonetheless, a survey conducted of 27 member institutions of the American Association of Colleges and Universities found that of the 150,072 undergraduate and graduate students participating 3.7 percent were gay/lesbian and 0.9 percent were transgender, genderqueer, or nonconforming.[64]

A question naturally arises as to the degree of responsiveness that colleges and universities have exhibited in relation to the increased diversity of the student body. While structural diversity is a necessary condition for attaining the educational benefits of diversity, it is not a sufficient one.[65] "Magical thinking" accompanies a belief that structural diversity automatically will bring about the educational benefits of diversity.[66] Many institutions have adopted token approaches to diversity and inclusion and have not implemented holistic approaches that take into account the deep and pervasive symbols, norms, and historical context that constitute an institution's identity.[67]

On predominantly white campuses, minoritized students often experience greater cultural dissonance with the policies and practices of a normative white culture and may seek ethnic subcultures or programs on their campuses.[68] For example, a qualitative study of five predominantly white campuses in three different geographical locations with 278 students found institutional negligence in diversity educational processes.[69] The researchers identified nine prevalent themes across these campuses including the fact that race was considered a taboo and avoidable topic across campus, except in ethnic studies classes; few spaces on campus had shared cultural ownership; and most students admitted to having few friends from differing racial and ethnic backgrounds.[70] From this vantage point, administrators need to understand the role of institutional culture in shaping student behaviors and experiences in order to create safe and comfortable environments that foster cross-cultural engagement across racial, ethnic, socioeconomic, gender, sexual orientation, and religious differences.[71]

Recently, however, several campuses have responded to student demands for the creation of affirming and supportive living spaces for minoritized students on predominantly white campuses. For example, the University of Connecticut (UConn) and the University of Iowa created living-learning communities for African American male students.[72] At UConn, the Schola2RS House (Scholastic House of Leaders who are African American Researchers & Scholars) is open to males of all races, but is designed principally to offer academic and social support, mentorship, and professional development to African American undergraduates who comprise a small percentage of the overall student population.[73] While critics such as Richard Sander and Stuart Taylor, Jr. have assailed such living arrangements as "clannish" and as racialized enclaves,[74] scholars such as Beverly Tatum have underscored the importance of minoritized student identity development as individuals respond to stereotypes arising from the dominant culture of white, male, heterosexual privilege and integrate and affirm their own positive identities.[75]

Promising Leadership Practices in Activating Diversity Learning

In the potent and conflicting cultural mix of national politics and student activism, we offer examples of institutional leaders who are making concrete diversity progress through tangible actions. Consider, for example, how President Robert Nelsen of Sacramento State University ("Sac State"), a public master's university with nearly 30,000 students, has realigned budget priorities to address what he describes as a "moral decision about diversity" and its significance. Nelsen, a white male, underscores the need to establish a sense of urgency about diversity and to articulate why diversity is important to student success. Although Sac State is the seventh most diverse institution west of the Mississippi and a majority minority institution, Nelsen indicates that the university is not yet the seventh most inclusive:

> Our university is sometimes too content with where we're at. There isn't a sense of urgency, urgency to bring diversity issues to the forefront and to have people say that this is important. That's been the biggest obstacle that we face. If it doesn't directly affect you, you tend not to do anything about it. We've seen that at the university. Meanwhile on the outside, we have people asking, 'Why are you wasting money and efforts on this? Your job is to graduate students, not to make them feel like they belong at the university.' They don't understand the connection between belonging and graduating: there is a direct, deep connection there. So unfortunately there's a modicum of resistance at the university, and there are outside politics surrounding discussions regarding inclusion and diversity. In all too many cases, it comes down to resources.
>
> . . . in our budget this year we set three priorities: priority number one was to make certain that we had enough classes for our students; priority two was diversity; priority three was safety. When you move diversity to the forefront of allocations, it means that with your budget decisions you are making a moral decision about diversity and its importance. That takes some explaining and teaching so that people understand why it is important.

In another example, at the University of Pittsburgh, Chancellor Patrick Gallagher has initiated an institutional diversity agenda that includes important structural steps through creating a diversity office in the Chancellor's office and identifying "embracing diversity and inclusion" as a goal of the strategic plan. Provost Patricia Beeson designated the 2016–2017 year as the Year of Diversity for the entire university and its regional campuses. The university has moved to institutionalize diversity within the cultural fabric of the university as exemplified by a key driver in the 2016–2020 strategic plan, "shaping the culture." The plan calls for a culture of investment in its people and the need to become more diverse and interconnected.[76] As Mario Browne, Director of the Office of Health Sciences Diversity, explains:

> The university from undergraduate on up through graduate programs would like to see more intentional diversity and inclusion in curriculum and pedagogy. The theme for this coming year will be diversity. That comes really from the Chancellor; one of his formal, overarching goals is diversity and inclusion, specifically creating a campus climate of that nature and also infusing diversity within the curriculum. So it's really coming from the top. The Provost has taken this charge up; there are actually some initiatives that the Provost's office is going campuswide.

Browne, an African American male, describes the collaborative efforts of all units in partnership with institutional leadership:

> we will have a year of diversity where all units will be encouraged to creatively think of ways to highlight what diversity and inclusion means reflective of their unit or reflective of their disciplines. And just get us all thinking about what that means collectively and individually. The Provost's office is a wonderful partner to have when trying to instill principles of diversity and inclusion in the curriculum across campus. The fact that there is an institutional commitment. . . . I think it [the year of diversity] will be very dynamic.

Given these promising leadership approaches, we begin the study by sharing the key research questions and methodology that guided our work and provide a brief overview of the organization of the book.

Research Focus of the Study

In emphasizing the centrality of a diversity culture shift as a major objective for institutions of higher education in the twenty-first century, our goal is eminently practical. We seek to address the following research questions within the framework of predominantly white public and private four-year colleges, master's level institutions, and doctoral research universities:

> What best practices offer the greatest potential for progress in diversity organizational learning?
>
> How do external and internal political pressures affect the ability of colleges and universities to develop diversity organizational learning strategies?
>
> In the face of overt and covert opposition, what leadership strategies, policies, structures, resources, and institutional practices have succeeded in moving the diversity organizational learning agenda forward?
>
> What common barriers within the higher education environment preclude comprehensive diversity culture change?
>
> What are the prevailing methods and approaches used for diversity professional development for faculty, administrators, staff and students in these institutions?
>
> What are the shortcomings/strengths associated with these methods?
>
> What methodologies, strategies, and practical approaches would address these shortcomings and build on existing strengths?

Organization of the Book

This book is organized thematically with the goal of assisting academic and administrative leaders in identifying leading-edge strategies for strengthening diversity organizational learning. We define leadership broadly to include both academic and administrative leaders whether at the cabinet level, in governance structures, at the school, college, or department level, or as faculty and staff advocates for diversity. We present five case studies of public and private research universities that illustrate the contextual challenges of implementing systematic programs of diversity organizational learning.

- In order not to overload the introductory chapter with dense definitions, Chapter 2 lays the groundwork for the study by delving into a discussion of the meaning and implications of diversity, inclusion, and organizational learning. It discusses the challenges of the campus landscape for diversity and provides an overview of twenty-first-century leadership and the primary roles responsible for leading a diversity culture shift. It then explores the evolution of the chief diversity officer role and the increasing role of tenured faculty and faculty governance as a force for diversity change. The chapter concludes with a case study of the University of Tennessee at Knoxville (UTK) that examines the impact of the state legislature defunding the Office for Diversity and Inclusion for fiscal year 2016–2017.

- Chapter 3 focuses on the Inclusive Excellence (IE) change model as the foundational approach that underpins systematic diversity organizational learning. The chapter explores research-based theories of cultural change and identifies hallmarks and criteria that delineate the predominant phases of systematic diversity organizational learning. The chapter includes a case study delineating the progressive course of diversity organizational learning at Princeton University.
- Chapter 4 offers representative approaches to diversity organizational learning based on research findings and interviews with diversity officers. The chapter identifies key challenges of diversity officers in implementing diversity education and their evaluation of the strengths and weaknesses of prevailing approaches. It shares common institutional barriers that preclude the development of systematic diversity education programs as well as concrete recommendations shared by diversity officers for developing successful approaches. The chapter also addresses specific platforms for differentiated organizational learning based upon stakeholder groups including administrators, faculty, staff, and students. It concludes by sharing a progressive taxonomy for diversity organizational learning. Two case studies are presented at the close of the chapter: Lehigh University and the University of Maryland at College Park.

- Chapter 5 summarizes key findings of the book and provides 20 characteristics of an integrated, systematic diversity organizational learning strategy. Appendix A describes the demographics of the diversity officer sample and Appendix B offers a planning matrix of action items for the implementation of diversity learning programs by campus leadership.

Survey and Interview Process

This book draws upon a series of in-depth interviews held with university and college leaders including diversity officers, provosts and executive officers, faculty, administrators, and students. Obtaining a representative sample of diversity officers for the study proved more difficult than anticipated. Initial outreach to diversity officers requesting participation in an online survey instrument that relied upon LinkedIn and personalized email invitations, but provided limited results. The lack of response led us to revisit our approach. Whereas the online survey questions focused on tactical issues such as staffing, types of training offered, attendance at training, and training budgets, our subsequent interview model addressed more strategic diversity and organizational learning issues. Our principal method was to rely on introductions by key individuals within an institution or in-person contacts generated through conferences. We discuss the likely reasons for the reluctance of diversity officers to participate in the study in Chapter 2. By contrast, in developing the case studies as well as providing additional commentary, tenured faculty expressed greater willingness to participate and frequently referred us to additional contacts. A demographic breakdown of the diversity officers participating in the survey is provided in Appendix A.

We conclude this introductory chapter with a case study on the University of Missouri at Columbia, the institution that served as a bellwether for the nation in fall 2015. The case study illustrates how courageous and unflinching student activism provided the spark that led to national protests and galvanized system and university administration, faculty, and students to accelerate significant structural change and the course of diversity organizational learning.

Case Study I
The Roller Coaster of Change at the University of Missouri at Columbia ("Mizzou")

This case study examines the confluence between complex factors representative of the forces at play in diversity change in public institutions of higher education across the nation. In late 2015, national events relating to the policing of African American citizens in Ferguson, Missouri, triggered student protests and unrest on the nearby flagship campus of the University of Missouri ("Mizzou"). The longstanding concerns of minoritized students regarding racial tensions and exclusionary treatment boiled over in protests at Mizzou and resulted in the formation of a student coalition called "Concerned Students 1950." The historical legacy of race relations in the state of Missouri was reflected in this activist group's choice of "1950" to commemorate the admission of the first African American students to the University of Missouri.

The uproar at Mizzou provides a microcosm for study of the development of diversity organizational learning. The case study illustrates the upheaval that can occur due to the lack of leadership responsiveness in addressing patterns of exclusion; the absence of structural diversity practices; and the failure to acknowledge and proactively address the prevailing racial climate for minoritized students. At the same time, the case study offers important practices for consideration in terms of the difference made by leadership commitment, the introduction of comprehensive diversity assessment efforts, the importance of structural and resource enhancements, and the pivotal role of tenured faculty leadership in the change process. As a vehicle for building meaningful interactions among faculty, staff, administrators, and students, the Faculty Council on Race Relations at Mizzou has pioneered a process for transformative diversity organizational learning through sustained small group interactions.

Historical Legacy

Both the state of Missouri and the University of Missouri at Columbia (UM or "Mizzou") have long been the focal point of racial tensions in the nation. Missouri was a border state between North and South during the Civil War, a state that permitted slavery but did not secede from the union. As one commentator notes, "the state suffers from some of the worst racial pathologies of both [the North and South] regions."[77]

Following the Civil War, the state constitution perpetuated a system of segregated education in which African American students could not attend the University of Missouri, but a separate university, Lincoln University, was established for these students. When Lloyd Gaines, an African American who was president of his all-African American undergraduate class at Lincoln University, filed a lawsuit for admission to Mizzou's law school rather than be sent out of state as was the practice, lawyers for the National Association for the Advancement of Colored People (N.A.A.C.P.) took the case to the Supreme Court in 1938. The high court ruled in *Gaines v. Canada* (S. Woodson Canada was the university registrar) that the state had failed to meet the "separate but equal" standard that had been established in the 1896 *Plessy v. Ferguson* case. As a result, the Court determined that Missouri needed to either admit Gaines or create a separate law school. The university chose the latter, but Gaines suddenly disappeared in Chicago in 1939.[78] Similarly, in 1939, Lucile Bluford, an African American woman, showed up on the flagship campus to register and was turned away because she was African American.[79] Bluford attempted 11 times to enter the University of Missouri and filed several law suits. When in 1941 the state Supreme Court ruled in her favor, the School of Journalism closed the school rather than admit her.[80]

In the fall of 1950, the University of Missouri accepted the first nine African American students and the first African American faculty member was hired in 1969.[81] The Legion of Black Collegians was founded in 1968, and, although not recognized by the university for several years, was devoted to changing the university environment in a peaceful manner.[82]

Even today the student population at Mizzou is not reflective of the demography of the state population. The university ranks the lowest among all 14 Southeastern Conference universities in the percentage of African American faculty.[83] According to Fall 2014 statistics, the University has an undergraduate enrollment of 27, 654 that is 79 percent white, 8 percent African American, 3 percent Hispanic, and 2 percent Asian American. The African American population in Missouri is approximately 12 percent and four public universities in the state including two historically African American institutions have greater percentages of African American student enrollment. Furthermore, the University has relatively few African American faculty members. Among the 1226 tenured/tenure-track faculty, 74.88 percent are white, 3.18 percent African American, and 15.91 percent Asian American.[84] Similarly, when taking into account the 900 part-time faculty, approximately three-fourths of the faculty are white. Hiring patterns over the last two years have not substantially increased the number of African American faculty. During this time period, the flagship campus hired 13 times more white faculty, with only 19 African American faculty among the 451 faculty hired.[85]

A Perfect Storm: Was Mizzou Paying Attention to Race Relations?

In December 2014, following the shooting of an African American citizen, Michael Brown, by a white police officer in Ferguson, Missouri, and nearly a year prior to the dramatic protests on the Mizzou campus, then Chancellor R. Bowen Loftin convened listening sessions to discuss race relations. The nearly week-long delay of the administration at Mizzou in responding to Ferguson had led to campus upheaval. Student Government President Payton Head shared his frustration with the bigotry and lack of inclusion on campus that made him feel unwelcome after individuals riding in the back of a pickup truck yelled racial slurs at him.[86]

Chancellor Loftin had come to Mizzou only a year earlier having served as president at Texas A&M University, an institution twice UM's size. In the words of one faculty member, Loftin "was shocked and authentically moved and disturbed" by what he heard at these sessions. Chancellor Loftin and Craig Roberts, Chair of Faculty Council and a white male professor of plant sciences, met with the Board of Curators to discuss the situation. It was determined that Faculty Council should form a committee on race relations. Journalism professor Berkeley Hudson, a white male from Mississippi, was asked to be the chair. Before assuming the role, Hudson asked the Chancellor, "Do you really care about this? Will you support me?"[87] Based on the affirmative response, Hudson agreed, despite the lack of remuneration for the additional assignment or course release from his regular teaching responsibilities.

Hudson spent four months interviewing people, attended several more listening sessions, and watched the Chancellor and other members of the administration "learning on their feet" in sometimes painful and awkward interactions with students. He wanted to find people of different backgrounds and beliefs for the 12-member committee that included Jonathan Butler, a graduate student who later went on a hunger strike. Michael Middleton, then deputy chancellor and law professor, was also on the committee.

The committee met throughout the summer and faced, in Hudson's words, "the intractable goal of naming the problem of race relations and coming up with solutions to problems." Hudson explains that in the intense, confidential meetings the members built trust and learned how to listen to each other and tell each other's stories. They sought to expand their points of view and not see things in binary ways. Taking note of the real structural problems in terms of diversity and inclusion on campus, Hudson looked forward to the replication of the committee structure and format in similar groups across campus in the future.

On October 10, 2015, students protesting racism at the homecoming parade blocked Missouri system president Tim Wolfe's car to express their concerns. Wolfe, an ex-software executive and corporate outsider hired by the system Board of Curators in 2012, did not respond or get out of his car and was even perceived to be smiling.[88] Moreover, Wolfe did not address the homecoming parade incident for over a month.[89] In a later encounter, Wolfe responded to students' requests for him to define "systematic oppression" by stating, "Systematic oppression is because you don't believe that you have the equal opportunity for success." When students expressed their disbelief that he would attribute such oppression to their perceptions, he turned his back and walked away.[90]

On October 20, Concerned Students 1950, a group of 11 African American men and women, issued seven key demands for structural changes related to diversity and inclusion. A member of the group, Jonathan Butler, launched a hunger strike on November 2 in response to both the university's intransigence on issues of inequality. Butler, who had been at Mizzou for seven years and experienced bullying and being attacked with racial slurs, signed a courageous "Do Not Resuscitate Order."[91]

In the dramatic events that followed, the linchpin was the football team's ultimatum. The team asked the system president, Timothy Wolfe, to resign or they would not play and would forfeit a $1 million game with Brigham Young University. On November 9, 2015, the eighth day of Butler's hunger strike, the Board of Curators held a special meeting and President Wolfe announced his resignation. Responding to the resignation, Jonathan Butler explained his own motivation for the hunger strike:

> I believe God knew in my heart how committed I was. It's more than just racism going on. It's more than just sexism going on. It's more than just homophobia. It's all these inequalities. It takes each of us in this community, each of us . . . to fight against these injustices and do what's right by the people that are next to us, standing shoulder by shoulder. I want everybody to understand, that this is not about me; this is about our community. 'Cause you saw what we did here . . . this is not Jonathan Butler, this is

> a community effort. We stood up for ourselves, we stood up as students; we stood up as staff; we stood up as faculty.[92]

Wolfe and top administrators were viewed by one faculty member as "tone deaf" to what was happening on social media with a tendency to "respond and react" rather than be proactive. In his resignation speech, Wolfe himself acknowledged the failure to listen and respond to student concerns, but criticized the way in which change had occurred:[93]

> It is my belief we stopped listening to each other. We didn't respond or react. We got frustrated with each other, and we forced individuals like Jonathan Butler to take immediate action and unusual steps to effect change.
>
> This is not, I repeat not, the way change should come about. Change comes from listening, learning, caring and conversation. We have to respect each other enough to stop yelling at each other and start listening, and quit intimidating each other. . . . Unfortunately this has not happened.

Chancellor Loftin resigned the same day as President Wolfe to be effective at the end of the year when he would retreat to a faculty position. His resignation was a result of other campus issues including pressure from nine deans resulting from his allegedly forcing the Vice Chancellor for Health Sciences to resign, revoking graduate students' health insurance and then restoring it, and a dispute over Planned Parenthood.

Contentious internal politics were revealed in an inflammatory email leaked after Wolfe's resignation, Wolfe accused Loftin of shifting the focus of Concerned Students 1950 to him "once he discovered his job was in jeopardy in late September."[94] He further chided the Board of Curators for appointing Michael Middleton, an African American who was former deputy chancellor and law professor, as interim president of the four-campus system. He accused Middleton of having "failed miserable [sic] in his capacity as the long time leader on diversity issues."[95]

The State Political Environment

Amidst the national headlines about student protests at Mizzou, the Republican-dominated state legislature reacted sharply to events on the flagship campus. Not only did criticism focus on how the campus handled the protests, but in particular about the way in which the university's governing board, the Board of Curators, was addressing the actions of Melissa Click. Click, an assistant professor of Communication, had worked at the campus since 2003, received a tenure-track appointment, and was recommended for tenure by the College of Arts and Sciences on November 9, 2015.[96] On that day, however, Click called for "muscle" in an attempt to push back two student reporters who were trying to interview the protestors. A "no media" sign had been posted at the rim of the campus quadrangle. Click had allegedly pushed the camera of one of the reporters.[97] Subsequently, Click was issued a letter of reprimand by the Provost, Garnett Stokes, and apologized for her behavior. Nonetheless, on January 4, 2016, over 100 legislators signed two letters calling for Click's termination.[98] Pressure continued from the legislature and on January 27, 2016 Click was suspended by the Board of Curators and subsequently terminated on February 24, 2016. She was not given the due process accorded as part of faculty governance in the university's rules and regulations.[99]

With the threat of cuts in budgetary appropriations looming large, some legislative efforts were devoted to trying to trim Mizzou's budget. When the dust settled in the budget process for 2016–2017, however, the legislature voted for a 4 percent increase in state funding for all colleges and universities and a 3.8 million cut for University of Missouri system administration.[100] Michael Middleton subsequently announced that 20 positions at the system level would be cut, 17 of which were vacant, and merit salary increases would be put on hold.[101]

Nonetheless, contentious political rhetoric about the University of Missouri system continued in light of the impending fall 2016 gubernatorial and attorney general elections. The Senate president stated that the legislature would not confirm any Board of Curator appointments made by outgoing democratic governor, Jay Nixon, until the new governor was in office. The nine-member board previously had only two African American curators who resigned in the wake of the student protests and Nixon appointed Mary Nelson, General Counsel for the St. Louis Community College District, as the sole remaining African American curator and only the second female curator.[102]

The Republican critique of the Mizzou campus continued unabated. For example, former U.S. attorney Catherine Hanaway, a Republican contender for the gubernatorial nomination, expressed her belief that administrative positions should be cut and professors should do more teaching.[103] Pressure from the public also came through phone calls, emails, and social media. Vice Chancellor for Advancement, Tom Hiles, reported 3400 negative contacts regarding the issues at Mizzou, calling it "the most trying five months" of his professional career with a drop in donations and pledges of $5 million in the immediate aftermath of the protests.[104]

Building a Collective Vision for Diversity Transformation

Despite such legislative and public pressures, interim system president, Michael Middleton calmly described the situation at a National Press Club luncheon in June 2016 and emphasized that Mizzou's issues regarding race relations were not unique:

> But the fact of the matter was that institutionally leaders were not paying attention and they got caught. Our students were concerned with race relations on our campus and felt extremely marginalized. . . . institutionally we were not listening to these very intelligent students, passionate students, who were telling leaders to wake up. What happened at MU I hope provides an

> instructional experience, a learning moment, a wakeup call for all who lead universities. And the lesson is that leaders must continually assess their campus climate to ensure that race relations are, in fact, good or excellent. Likely, they are not as good as you think.[105]

Middleton further added:

> The problems our university faces in this area are real. In fact, universities nationwide are facing these same problems, as is our nation as a whole.[106]

Middleton, 68, an African American male, had founded Mizzou's Legion of Black Collegians 46 years earlier, a group that had issued a remarkably similar set of demands as the student activist group, Concerned Students 1950.[107] The Black Collegians reported a "nonchalant attitude on the part of the university" despite physical threats against African Americans as well as the Black Culture House.[108] On accepting the interim appointment as president of the four-campus system, Michael Middleton had stated:

> The mission of our great university is to discover, disseminate, preserve, and apply knowledge. To this end we must confront many uncomfortable societal issues, that once confronted will make us stronger.[109]

Middleton began a process of collective visioning with the four campus chancellors, captured in a video titled "We Imagine." The video featured the voices of faculty, staff, students, and administrators who envisioned the university as an inclusive institution in which all are valued and differences are celebrated.[110] Middleton himself articulated this collective vision as follows:

> our collective vision of the university we all imagine: One that is relentless in the pursuit of knowledge, freedom and justice that liberates and uplifts us all. One where we can all live and learn together, respectfully and responsibly. One anchored in

> the greater community. One where creativity, innovation, and courage are hallmarks of our success. The future of our university is not shaped by our vision alone, but yours as well.[111]

Changes initiated at the University of Missouri system level included the hiring of the inaugural system chief diversity officer, Kevin McDonald, who simultaneously was appointed as interim vice chancellor for inclusion, diversity, and equity at Mizzou. Other new systemwide initiatives included the formation of a diversity and equity task force and the hiring of a consultant to undertake a diversity audit designed to provide an assessment of diversity programs and processes and an analysis of how policies, leadership, and culture support or serve as barriers to diversity and inclusion.[112] In addition, funds of $921,000 were allocated to conduct a campus climate survey, provide additional training on diversity, and support students' mental health.[113]

Campus Shakeup and the Impact on Diversity Progress

Dr. Henry C. "Hank" Foley, a white male who was formerly the senior vice chancellor for research and graduate studies and UM system executive vice president for academic affairs, research, and economic development, was named interim chancellor of the Mizzou campus. A number of interim appointments and personnel changes were made under Foley's direction. A new position of Vice Chancellor (VC) for Inclusion and Equity was created and held on an interim basis by Chuck Henson, an African American associate law school dean. Subsequently, with the appointment of Kevin McDonald, Mizzou created a new division titled the Division of Inclusion, Diversity, and Equity with a budget of $1.5 million. Five student support centers and the offices handling civil rights and ADA compliance were placed under the division's umbrella.[114]

High turnover among administrators included the departure through resignation or retirement of six deans as well as the departure of the athletic director, Mark Rhoades, after only 14 months. The administrative shakeup left the university administration and athletic department "barely recognizable" in comparison to the prior fiscal year.[115] In light of the significant turnover and plethora of interim appointments, Jon, a white male faculty member indicated that decisions on difficult matters were likely to be put on hold:

> Nobody's going to make the hard choices. . . . We are just doing the stuff that doesn't require anybody to make the hard choices. . . . I think it is a similar thing in the diversity world which is: "Okay, we'll fund this guy. You want a diversity guy, here he is." But are we going to really engage in difficult conversations about what we want to do in faculty hiring? Everyone now has to go through some sort of training on hiring. That's an initiative that's gotten people's time and attention. But the bigger stuff, the stuff that really depends upon joint effort from the Chancellor and Provost to push the campus in a certain way, that's sustained and focused? Those things are extremely difficult right now.

In his "State of the University" address delivered on January 27, 2016, Chancellor Foley, a physical chemist, emphasized the need for academic values including academic freedom to pursue research and scholarship, institutional autonomy, and shared governance. He referred to external, environmental pressures for change as well as to Stephen Jay Gould's theory of "punctuated equilibrium" in which longer periods of incremental, evolutionary change are punctuated with rapid periods of revolutionary change.[116] Foley emphasized the need to adapt quickly and for the university to keep pace with change and be more fully inclusive. He repeatedly emphasized the university's role in addressing the expectations of "all students."[117]

While much of the State of the University speech was devoted to noting specific university, faculty, staff, and student accomplishments, towards the end of the speech Foley did indicate that matters of diversity and social justice were a priority. At the close of the speech, Foley stressed the importance of changing the culture to one of "higher respect" and "behaving differently" to make new strategies and initiatives effective.[118] Yet when listing priority goals at the end of the speech, diversity and inclusion were not mentioned.

Consistent with the omission of diversity as a strategic goal in Chancellor Foley's State of the University speech, Lisa, a tenured white female faculty member, was skeptical as to the prioritization of a systemic course related to diversity and organizational learning. As she indicates,

> His messages aren't just about diversity . . . he is sending a lot of messages. I don't think that there is necessarily a clear, consistent, "Let's put this out on the forefront: let's think about how organizations transform." I think hiring a VC for diversity and inclusion is incredibly important, but I hope that people don't just rest on the laurels of this person fixing the whole situation. Because that's not going to happen. It's too big a job for one person. And maybe Kevin at the VC level will be able to have more influence on how we think about ourselves as a learning organization and how diversity is central to that learning organization. The other thing that complicates this is our budget situation. As you well know, often some of the first things to be taken away are around diversity when there is a budget crisis.

It is unclear as whether or not the student demonstrations will lead to sustained structural and cultural change and whether such changes will be pervasive rather than symbolic. African American students at Mizzou have reported that racial tensions are a function of everyday life on campus and threats against African American students have been posted on the social media site, Yik Yak.[119] Minoritized students have reported feeling unsafe and Law School faculty member and campus Faculty Council chair, Ben

Trachtenberg, a white male, noted that African American students had reported the use of the "N" word and were facing a hostile environment. In his words, "the atmosphere on this campus needs to change. And that's not easy to do. I am not going to come forward and say I have a magic bullet. But we need to work together to educate people."[120]

Further, as Lisa explains, the campus has gone through similar cycles of protest in the past and hopes that the students who are returning after the summer break will keep the issue of diversity on the front burner:

> I think it [the protests] served as the most prominent reminder that this is something that we should be focusing on and that as an organization we need to make this part of our culture. However, the long-term impact, I don't know. This is not the first time that there have been protests. It was perhaps the most salient and the most visible in large part of social media and because of our 24 hour news cycle. There have been protests over the years here; this is not a new topic for our campus. It's kind of like a roller coaster: we go up for a while and then we go down, and we go up for a while and then we go down. I would not be surprised if we see this cycle continue, and I'm hopeful that when students are back on campus that they remind the campus that this issue still needs to be addressed.

A number of factors contribute to the unpredictability of sustained attention to cultural change and diversity learning. With the intervening summer months and the lack of student presence, the momentum around diversity change has subsided. Concerned Students 1950, the group of 11 African American students who had issued seven key demands for change, disbanded amid internal tensions. Among the students' demands were the need for comprehensive racial awareness and inclusion curriculum offered by all campus departments, with mandatory training for faculty, administrators, staff, and students and a curriculum overseen by a board of faculty, staff, and students of color.[121] Furthermore, Jonathan Butler, the master's degree student who conducted the eight-day hunger strike, graduated as did Payton Head, the student government president.

Diversity Organizational Learning: The Mizzou Miracle?

Prior to his resignation, Chancellor Loftin who had been accused of not fighting racism on campus, instigated mandatory diversity training for faculty, staff, and students. While some hailed this step, others called it a "Band-Aid," "meaningless," and even "patronizing."[122] During Chuck Henson's brief tenure as interim VC for Inclusion, Diversity, and Equity, new organizational learning initiatives were launched including the "Diversity@Mizzou" orientation training program required for incoming students, search committee training, new professional development offerings, and guest lectures.[123] A mandatory citizenship training was introduced for new students in fall 2016, titled Citizenship@Mizzou. Stephanie Shonekan, chair of the Department of Black Studies who helped design the program said that the students and faculty she consulted preferred the word "citizenship" to "diversity" to address the values that make good citizens of Mizzou.[124]

Chancellor Foley indicated he wanted diversity training to be scaled up for all faculty, administrators, and staff, and also reported that the Chancellor's and Provost's staff received training on implicit bias from an outside facilitator. Although the implicit bias training was presented from a research-based perspective, such training needs to not only address individual prejudice but also build awareness of systemic social and institutional patterns of racialization.[125]

Faculty training, other than search committee training, remains a work in progress. A four-week "Diversity 101" course is offered online that can be taken by individuals or groups and requires three to five hours a week of time.[126] It involves weekly and content and discussion posts as well as interactive responses to posts. Billed as "non-threatening and inclusive," the training is voluntary and includes videos, a reflection journal, and self-guided activities.[127] Since the training is not fine-tuned to different groups, it requires a significant time commitment and appears to be primarily focused on raising awareness in a non-threatening manner, its impact on campus culture will likely be very limited.

A three-credit hour diversity course requirement was passed for the largest college on campus, the College of Arts and Sciences, following the creation of a faculty committee under the leadership of Angela Speck, professor of Astrophysics.[128] In addition, interim Chancellor Foley announced the use of $1 million from the Intellectual Property Revenue fund to be directed toward the recruitment of minority postdoctoral fellows who would be named as pre-faculty fellows with the goal of retaining these individuals in the future.[129]

The Faculty Council on Race Relations issued a report on September 15, 2016 summarizing its findings and stating, "We are talking about a cultural shift, a Mizzou Miracle as it were."[130] The report recommended the creation of small groups of individuals who are committed both to naming issues of race relations but also identifying solutions.[131] It also recommended an audio and video project to capture stories of race relations within the Mizzou community. The committee already has developed a series of videos that examine the racial landscape, describe ways to change it, and respond to skeptics of racism.[132] In addition, it will be issuing a series of podcasts primarily aimed at majority faculty.[133]

In terms of the recommendation to replicate the Council's format through small committees across campus, Ben Trachtenberg, Chair of Faculty Council, is unsure whether the work will receive the time and attention it needs. As he explains:

> I think the Race Relations Committee is doing important work. It is a very difficult question as to whether it can scale. I think it's been very useful for people in the room. It has been particularly

> useful as an educational time for white people in the room. I think the argument could be made that is somewhat taxing for the black people in the room. They're spending a lot of time talking about their strife to educate the white faculty. . . . we're asking the students to talk about things like racial slurs and things like that. . . . But are we going to assemble a bunch of committees like that? . . . are people going to sign up? Are they going to meet? I think they have a dozen members, and they . . . did a lot of work to get people on that committee, including Mike Middleton. . . . how many folks like that are available? . . . I don't know. And so I am very interested to see what happens with efforts to roll this out in the coming year.

Berkeley Hudson sees the experiences at Mizzou can serve as a turning point, a cultural shift, and a learning experience for institutions of higher education as a whole. As he asks:[134]

> And I ask provocative questions: What can each of us do today to make Mizzou a safe and welcoming place? What can we do to make Mizzou a local, national and global leader in race relations in terms of teaching, service, research and economic development?

The impact of deepened diversity learning on student experiences is reflected in the voices of students in Hudson's capstone Advanced Writing journalism course who wrote about race relations at Mizzou as part of course requirements.[135] For example, as Thom Dixon, a white male, noted,

> You don't have to think about it [race] every day. You don't have to think about it. But that's not true for people of different skin colors.

Similarly, Max Havey, a white male, described his own growing awareness of the existence of privilege:

> I don't know if it was biases, but more kind of privilege, not realizing it was there. Realizing the experience of black students

> on campus was not something I had really thought about or pondered or really put a lot of thought into. I kind of lived off in my own little bubble, my own little world . . . I knew I was coming from a background of privilege.

From a leadership standpoint, the turbulent student protests that led to the resignation of Timothy Wolfe as system president of the Missouri system were followed by the selection of the first Asian American president, Mun Choi, in late 2016. Choi, a seasoned academic with a doctorate in mechanical and space engineering from Princeton University, had served as the University of Connecticut's Provost since December 2012, Choi immediately expressed his desire for the development of a collective vision, stating, "The voices of faculty, students and staff—the heart and soul of the institution—matter greatly to me."[136] With changes in system leadership, the creation of a systemwide diversity position, the implementation of a systemwide campus climate study, and reinvigorated diversity efforts at the University of Missouri at Columbia, a new chapter is underway in the university's history. Interim chancellor Hank Foley left UM in May 2017 to become president of the New York Institute of Technology. He was succeeded by Alexander Cartwright, former provost and executive vice chancellor in the State University of New York System.

Looking ahead, Kevin McDonald, the system's inaugural chief diversity officer and interim vice chancellor of inclusion, diversity, and equity for the Mizzou campus, indicates that he is very optimistic that Mizzou is well on its way to changing the narrative regarding race relations. His leadership focus is on strengthening transparency and creating a positive campus climate that helps faculty, staff, and students feel safe and engaged and builds a strong sense of community. With an emphasis on the organic development of diversity initiatives and strategic planning, McDonald seeks to draw on community input to create a message and programmatic efforts around diversity that resonate with stakeholders and have an impact on the issues of concern for different constituencies. This approach, in his view, will foster continued diversity progress and also appeal to both urban and rural communities in the surrounding areas in Missouri.

Notes

1. Frenk, J. (2016, May 15). Why we need a 'scholarship of belonging.' *The Chronicle of Higher Education*. Retrieved August 9, 2016, from http://chronicle.com/article/Why-We-Need-a-Scholarship-/236443
2. Graham, D. A. (2016, October 12). The Joshua generation: Did Barack Obama fulfill his promise? *The Atlantic*. Retrieved January 12, 2017, from www.theatlantic.com/politics/archive/2016/10/obama-greensboro-clinton-joshua-generation/503615/
3. Ibid.
4. Chang, J. (2016). *We gon' be alright: Notes on race and resegregation*. New York: Picador.
5. Ibid., p. 40
6. Kezar, A. (2014). *How colleges change: Understanding, leading, and enacting change*. New York: Routledge.
7. Ibid., para 5.
8. Prasad, A. (2001). Understanding workplace empowerment as inclusion. *The Journal of Applied Behavioral Science*, 37(1), 51–69.
9. Feagin. (2006). *Systemic racism: A theory of oppression*. New York: Routledge.
10. Ragins, B. R. (2007). Diversity and workplace mentoring relationships: A review and positive social capital approach. In T. D. Allen and L. T. Eby (Eds.), *The Blackwell handbook of mentoring: A multiple perspectives approach* (pp. 281–300). Malden, MA: Blackwell Publishing.
11. The term "minoritized" is used instead of "minority" to reference how the socially constructed reproduction of inequality subordinates individuals from underrepresented racial/ethnic groups as minorities within predominantly white campus environments. See Harper, S. R. (2012). Race without racism: How higher education researchers minimize racist institutional norms. *Review of Higher Education*, 36(1), 9–29.
12. Kezar, A. J. (2007). Tools for a time and place: Phased leadership strategies to institutionalize a diversity agenda. *The Review of Higher Education*, 30(4), 413–439.
13. Jaschik, S. (2014). Students protest over Ferguson: Many voice outrage over lack of an indictment. *Inside Higher Ed*. Retrieved April 21, 2017, from www.insidehighered.com/news/2014/11/25/students-protest-lack-indictment-ferguson
14. Hartocollis, A., and Bidgood, J. (2015, November 11). Racial discrimination protests ignite at colleges across the U.S. *The New York Times*. Retrieved April 2, 2015, from www.nytimes.com/2015/11/12/us/racial-discrimination-protests-ignite-at-colleges-across-the-us.html
15. Chessman, H., and Wayt, L. (2016). What are students demanding? *Higher Education Today*. Retrieved August 9, 2016, from https://higheredtoday.org/2016/01/13/what-are-students-demanding/
16. Ibid.

17. Dickey, J. (2016, May 31). The revolution on America's campuses. *Time Magazine*. Retrieved August 28, 2016, from http://time.com/4347099/college-campus-protests/
18. Hartocollis, A. (2016, August 4). College students protest, alumni's fondness fades and checks shrink. *The New York Times*. Retrieved August 28, 2016, from www.nytimes.com/2016/08/05/us/college-protests-alumni-donations.html?_r=0
19. Ibid.
20. Concerned Graduate Students of Color. (2015, November 16). Graduate solidarity statement and demands. *Bluestockings Magazine*. Retrieved August 28, 2016, from http://bluestockingsmag.com/2015/11/16/graduate-solidarity-statement-and-demands/
21. Simundich, J. (2015, November 26). Why Brown University's $100 million plan to improve race relations falls short. *The Huffington Post*. Retrieved August 22, 2016, from www.huffingtonpost.com/joel-simundich/brown-u-plans-to-spend-10_b_8650776.html
22. *Pathways to diversity and inclusion: An action plan for Brown University*. (2016). Retrieved August 28, 2016, from www.brown.edu/about/administration/institutional-diversity/pathways
23. Yeung, F. P. F. (2013). Struggles for professional and intellectual legitimacy: Experiences of Asian and Asian American female faculty members. In S. D. Museus, D. C. Maramba, and R. T. Teranishi (Eds.), *The misrepresented minority: New insights on Asian Americans and Pacific Islanders, and the implications for higher education* (pp. 281–293). Sterling, VA: Stylus.
24. Ibid.
25. Yan, W., and Museus, S. D. (2013). Asian American and Pacific Islander faculty and the glass ceiling in the academy: Findings from the national study of postsecondary faculty. In S. D. Museus, D. C. Maramba, and R. T. Teranishi (Eds.), *The misrepresented minority: New insights on Asian Americans and Pacific Islanders, and the implications for higher education* (pp. 249–265). Sterling, VA: Stylus.
26. Wu, F. (2017, January 24). Personal communication.
27. *Yale launches five-year, $50 million initiative to increase faculty diversity*. (2015). Yale University. Retrieved August 30, 2016, from http://news.yale.edu/2015/11/03/yale-launches-five-year-50-million-initiative-increase-faculty-diversity
28. *Report on faculty diversity and inclusivity in FAS*. (2016), p. 5. Retrieved August 30, 2016, from http://fassenate.yale.edu/sites/default/files/files/Reports/FAS%20Senate%20-%202016-05-19%20-%20Diversity%20and%20InclusivityFINAL%20copy%202.pdf
29. Ibid., p. 7.
30. *Key initiatives*. (2016). Yale University. Retrieved August 30, 2016, from http://inclusive.yale.edu/key-initiatives
31. *Diversity, equity & inclusion: Strategic plan*. (2016). University of Michigan. Retrieved February 22, 2017, from https://diversity.umich.edu/wp-content/uploads/2016/10/strategic-plan.pdf
32. Kabbany, J. (2016, October 28). Report: University of Michigan's new diversity chief to earn $385,000 annually. *The College Fix*. Retrieved February 23, 2017, from www.thecollegefix.com/post/29694/
33. *Diversity, equity & inclusion: Strategic plan*. (2016). University of Michigan.
34. Reid, D. (2016, September 19). University names first chief organizational learning officer. *The University Record*. Retrieved February 22, 2017, from https://record.umich.edu/articles/university-names-first-chief-organizational-learning-officer
35. Ibid.
36. Levison, A. (2016, March 8). Will Trump send working-class whites to the democrats? *The New York Times*. Retrieved August 9, 2016, from www.nytimes.com/2016/03/08/

opinion/campaign-stops/will-trump-send-working-class-whites-to-the-democrats.html?_r=1

37. Anderson, C. (2016). *White rage: The unspoken truth of our racial divide*. New York: Bloomsbury.
38. Ibid.
39. Feagin, J. R. (2012). *White party, white government: Race, class, and U.S. politics*. New York: Routledge.
40. Anderson. (2016). *White rage*, p. 5.
41. Feagin, J. R., and Ducey, K. (2017). *Elite white men ruling: Who, what, when, where, and how*. New York: Routledge.
42. Confessore, N. (2016, July 13). For whites sensing decline, Donald Trump unleashes words of resistance. *The New York Times*. Retrieved August 10, 2016, from www.nytimes.com/2016/07/14/us/politics/donald-trump-white-identity.html?_r=1
43. Ibid. See also Chang. (2016). *We gon' be alright*, p. 2.
44. Cillizza, C. (2015, August 27). Donald Trump likes to talk about the 'silent majority.' For many, that has racial overtones. *The Washington Post*. Retrieved August 10, 2016, from www.washingtonpost.com/news/the-fix/wp/2015/08/27/donald-trump-keeps-talking-about-the-silent-majority-is-that-a-racial-dog-whistle/
45. Young, A. (2016, March 14). NAACP president: Trump 'kind of Jim Crow with hairspray and a blue suit.' *CNN Politics*. Retrieved August 10, 2016, from www.cnn.com/2016/03/14/politics/donald-trump-naacp-cornell-william-brooks/
46. Giroux, H. A. (2016). *Trump's popularity and the politics of apology and racial cleansing*. Retrieved August 10, 2016, from www.academia.edu/23430442/Trump_s_Popularity_and_the_Politics_of_Apology_and_Racial_Cleansing?auto=download
47. Logue, J. (2016, March 15). Trump as a taunt. *Inside Higher Ed*. Retrieved August 10, 2016, from www.insidehighered.com/news/2016/03/15/trump-used-taunt-against-students-and-minority-groups?utm_source=Inside+Higher+Ed&utm_campaign=d6fcbeddde-DNU20160315&utm_medium=email&utm_term=0_1fcbc04421-d6fcbeddde-198511277#.VuiRz9hDamE.mailto
48. Ibid.
49. Ibid.
50. Vasquez, T. (2015). I've experienced a new level of racism since Donald Trump went after Latinos. *The Guardian*. Retrieved August 10, 2016, from www.theguardian.com/commentisfree/2015/sep/ 09/donald-trump-racism-increase-latinos
51. Southern Poverty Law Center. (2016). *The Trump effect: The impact of the presidential campaign on our nation's schools*. Retrieved August 10, 2016, from www.splcenter.org/sites/default/files/splc_the_trump_effect.pdf
52. Ibid.
53. Picca, L. H., and Feagin, J. R. (2007). *Two-faced racism: Whites in the backstage and frontstage*. New York: Routledge.
54. Ibid.
55. *White supremacists on campus: Unprecedented recruitment efforts underway*. (2017, March 6). Anti-Defamation League. Retrieved April 25, 2017, from www.adl.org/blog/white-supremacists-on-campus-unprecedented-recruitment-efforts-underway?_ga=1.3868912.1741098207.1488839158
56. Lowery, W., and Fletcher, M. A. (2015, November 22). The Michael Brown shooting changed my life. *The Washington Post*. Retrieved August 10, 2016, from www.washingtonpost.com/business/economy/the-michael-brown-shooting-changed-my-life/2015/11/22/4ad12b94-8bac-11e5-acff-673ae92ddd2b_story.html
57. Frey, W. H. (2015). *Diversity explosion: How new racial demographics are remaking America*. Washington, DC: Brookings Institution Press.
58. Ibid.

59. Eagleton-Pierce, M. (2016). *Neoliberalism: The key concepts.* New York: Routledge. See also Museus, S. D., Ledesma, M. C., and Parker, T. L. (2015). *Racism and racial equity in higher education* (ASHE-ERIC Higher Education Reports, Vol. 42, No. 1). San Francisco: Jossey-Bass.
60. Ahonen, P., Tienari, J., Merilainen, S., and Pullen, A. (2014). Hidden contexts and invisible power relations: A Foucauldian reading of diversity research. *Human Relations*, 67(3), 263–286.
61. Hussar, W. J., and Bailey, T. M. (2013). *Projections of education statistics to 2021* (40th ed.). Retrieved August 11, 2016, from http://nces.ed.gov/pubs2013/2013008.pdf
62. The data must be interpreted with caution due to the changes in 2010 census reporting that added a separate category for two or more races as well as for native Hawaiian/Pacific Islander.
63. Kuh, G. D., Kinzie, J., Buckley, J. A., Bridges, B. K., and Hayek, J. C. (2006). *What matters to student success: A review of the literature: Executive summary.* Retrieved August 13, 2016, from https://nces.ed.gov/npec/pdf/Kuh_Team_ExecSumm.pdf
See also Chun, E. (2013). Meet our nation's first-generation college students. *INSIGHT into Diversity*, 82(2), 22.
64. Cantor, D., Fisher, B., Chibnall, S., Townsend, R., Lee, H., Bruce, C, and Thomas, G. (2015). *Report on the AAU campus climate survey on sexual assault and sexual misconduct.* Retrieved August 16, 2016, from www.aau.edu/uploadedFiles/AAU_Publications/AAU_Reports/Sexual_Assault_Campus_Survey/Report%20on%20the%20AAU%20Campus%20Climate%20Survey%20on%20Sexual%20Assault%20and%20Sexual%20Misconduct.pdf
65. Gurin, P., Dey, E. L., Hurtado, S., and Gurin, G. (2002). Diversity and higher education: Theory and impact on educational outcomes. *Harvard Educational Review*, 72(3), 330–367.
66. Chang, M. J. (2007). Beyond artificial integration: Reimagining cross-racial interactions among undergraduates. *New Directions for Student Services*, 120, 25–37.
67. Jayakumar, U. M., and Museus, S. D. (2012). Mapping the intersection of campus cultures and equitable outcomes among racially diverse student populations. In S. D. Museus and U. M. Jayakumar (Eds.), *Creating campus cultures: Fostering success among racially diverse student populations* (pp. 1–27). New York: Routledge.
68. Ibid.
69. Harper, S. R., and Hurtado, S. (2007). Nine themes in campus racial climates and implications for institutional transformation. *New Directions for Student Services*, 120, 7–24.
70. Ibid.
71. Museus, S. D. (2008). Focusing on institutional fabric: Assessing campus cultures to enhance cross-cultural engagement. In S. R. Harper (Ed.), *Creating inclusive campus environments for cross-cultural learning and student engagement* (pp. 205–234). Washington, DC: National Association of Student Personnel Administrators.
72. New, J. (2016, February 2). For black men. *Inside Higher Ed.* Retrieved August 13, 2016, from www.insidehighered.com/news/2016/02/02/u-connecticut-creates-new-living-learning-center-black-male-students
73. Reitz, S. (2016, April 7). UConn to welcome students in two new learning communities. *UConn Today.* Retrieved August 13, 2016, from http://today.uconn.edu/2016/04/uconn-to-welcome-students-in-two-new-learning-communities/
74. Sander, R., and Taylor, S., Jr. (2012). *Mismatch: How affirmative action hurts students it's intended to help, and why universities won't admit it.* New York: Basic Books.
75. Tatum, B. D. (1997). *"Why are all the black kids sitting together in the cafeteria?": A psychologist explains the development of racial identity.* New York: Basic Books.

76. *The plan for Pitt: Making a difference together: Academic years 2016–2020.* (2015). Retrieved August 16, 2016, from www.pitt.edu/sites/default/files/ThePlanforPitt1023 2015.pdf
77. Marans, D., and Stewart, M. (2015, November 16). Why Missouri has become the heart of racial tension in America: From Ferguson to Mizzou, the show-me state is now a focal point. *The Huffington Post.* Retrieved August 19, 2016, from www.huffingtonpost.com/entry/ferguson-mizzou-missouri-racial-tension_us_ 564736e2e4b08cda3488f34d
78. Stout, D. (2009, July 11). A supreme triumph, then into the shadows. *The New York Times.* Retrieved August 19, 2016, www.nytimes.com/2009/07/12/us/12gaines.html?_r=1
79. Shahriari, S., and Smith, R. (2016). *Mizzou at a crossroads: Chapter one: Past and present.* Retrieved August 19, 2016, from http://apps.kbia.org/mizzou-crossroads/chapter-one-past-and-present.html
80. *Lucile H. Bluford.* (n.d.). Retrieved August 21, 2016, from http://shsmo.org/histori cmissourians/name/b/bluford/
81. Ibid.
82. Ibid.
83. Dennis, V., and Santhanam, L. (2015, November 10). It's true, Mizzou's black faculty numbers are low—Really low. *PBS Newshour.* Retrieved August 21, 2016, from www. pbs.org/newshour/updates/true-mizzous-black-faculty-numbers-low-really-low/
84. *Mizzou Diversity.* (2014). Retrieved July 3, 2016, from https://diversity.missouri.edu/ about/stats/students.php
85. Williams, M. R. (2016, June 4). University of Missouri campus in Columbia hires black faculty at a slow pace. *The Kansas City Star.* Retrieved August 19, 2016, from www. kansascity.com/news/state/missouri/article81848322.html
86. Pearson, M. (2015, November 10). A timeline of the University of Missouri protests. *CNN.* Retrieved August 19, 2016, from www.cnn.com/2015/11/09/us/missouri-protest-timeline
87. Hudson, B. (2016, July 5). Personal communication.
88. Pearson. (2015, November 10). A timeline of the University of Missouri protests.
89. Kingkade, T. (2015, November 10). The incident you have to see to understand why students wanted Mizzou's president to go: What happened at a homecoming parade set the stage for Tim Wolfe's eventual resignation. *The Huffington Post.* Retrieved August 21, 2016, from www.huffingtonpost.com/entry/tim-wolfe-homecoming-parade_us_56402 cc8e4b0307f2cadea10
90. Miller, M. E. (2015, November 9). With $1 million at stake, U. of Missouri's president now taking protests seriously. *The Washington Post.* Retrieved August 19, 2016, from www.washingtonpost.com/news/morning-mix/wp/2015/11/09/with-1-million-at-stake-u-of-missouris-president-now-taking-protests-seriously/
91. Chang. (2016). *We gon' be alright.*
92. Lee, S. (2016). *2 fists up.* Retrieved June 20, 2016, from www.youtube.com/watch? v=NLOG4EKeGOM
93. Vandelinder, E. (2015, November 9). Transcript of Tim Wolfe's resignation speech. *Missourian.* Retrieved August 19, 2016, from www.columbiamissourian.com/news/ higher_education/transcript-of-tim-wolfe-s-resignation-speech/article_b666ff76-8703-11e5-896f-87b19edd2aed.html
94. Jackson, A. (2016). *'This kind of attack was unexpected': Target of ex-Mizzou president's leaked email hits back.* Retrieved August 21, 2016, from www.businessinsider. com/bowen-loftin-hits-back-and-former-missouri-president-tim-wolfes-leaked-email-2016-1

95. Text of Tim Wolfe's letter criticizing University of Missouri leaders. (2016, January 27). *The Kansas City Star*. Retrieved August 21, 2016, from www.kansascity.com/news/local/article56868723.html
96. American Association of University Professors. (2016). *Academic freedom and tenure: University of Missouri (Columbia)*. Retrieved August 19, 2016, from www.aaup.org/file/UMColumbia_0.pdf
97. Ibid.
98. Ibid.
99. Ibid.
100. Keller, R. (2016, April 22). Boost for University of Missouri budget passes easily as legislative rancor ebbs. *Columbia Daily Tribune*. Retrieved August 19, 2016, from www.columbiatribune.com/news/education/turmoil_at_mu/boost-for-university-of-missouri-budget-passes-easily-as-legislative/article_357df4e6-d57f-5ca5-80ce-afe8ab79bec8.html
101. Nixon appoints 3 new curators to lead University of Missouri. (2016, June 8). *CBS St. Louis*. Retrieved August 21, 2016, from http://stlouis.cbslocal.com/2016/06/08/nixon-appoints-3-new-curators-to-lead-university-of-missouri/
102. Ibid.
103. Missouri governor hopefuls debate Mizzou, taxes, race relations. (2016, June 7). *CBS St. Louis*. Retrieved August 21, 2016, from http://stlouis.cbslocal.com/2016/06/07/missouri-governor-hopefuls-debate-mizzou-taxes-race-relations/
104. Keller, R. (2016, July 13). MU sets fundraising record despite campus turmoil, leadership turnover. *Columbia Daily Tribune*. Retrieved August 19, 2016, from www.columbiatribune.com/news/education/turmoil_at_mu/mu-sets-fundraising-record-despite-campus-turmoil-leadership-turnover/article_43eaae41-f5bb-5209–8cbf-0d9e1981e63d.html
105. Middleton, M. (2016, June 21). *University of Missouri president Michael Middleton speaks at the National Press Club*. Retrieved August 22, 2016, from www.youtube.com/watch?v=91XQwylcQAs&app=desktop
106. Ibid.
107. Salter, J., and Suhr, J. (2015, November 13). University of Missouri starts reviewing demands from student activists: The demands are strikingly similar to those from 1969. *The Huffington Post*. Retrieved August 22, 2016, from www.huffingtonpost.com/entry/university-of-missouri-demands_us_56468f35e4b0603773492b6e
108. Ibid.
109. University of Missouri names interim president. (2015). *CNN*. Retrieved August 22, 2016, from www.youtube.com/watch?v=ck9tpnLS-Ds&app=desktop
110. *University of Missouri System—We imagine*. (2016). Retrieved August 22, 2016, from www.youtube.com/watch?v=u-cHAvtm51c&feature=youtu.be&app=desktop
111. Middleton, M. (2016, June 23). Email communication.
112. *Diversity, equity and inclusion audit*. (2016). University of Missouri System. Retrieved August 20, 2016, from www.umsystem.edu/deiaudit
113. University of Missouri names interim president. (2015).
114. Keller, R. (2016, July 19). New university of Missouri division pulls together student centers, civil rights offices. *Columbia Daily Tribune*. Retrieved August 19, 2016, from http://m.columbiatribune.com/news/education/turmoil_at_mu/new-university-of-missouri-division-pulls-together-student-centers-civil/article_d816a016-ce73–5bd8-b822-cdf706bf8917.html
115. Walljasper, J., and Toppmeyer, B. (2016, July 14). After just 14 months, Rhoades leaves Missouri for Baylor. *Columbia Daily Tribune*. Retrieved August 23, 2016, from www.

columbiatribune.com/sports/rhoades-leaving-missouri-for-baylor/article_80f14566-8bca-5082-a564-014ce6f4666c.html
116. Shahriari and Smith. (2016). *Mizzou at a crossroads.*
117. Bold new Mizzou: Interim Chancellor Hank Foley calls for a better, stronger, more adaptable university. (2016, January 28). *Mizzou News.* Retrieved August 23, 2016, from https://news.missouri.edu/2016/bold-new-mizzou/. See also Shahriari and Smith. (2016). Mizzou at a crossroads.
118. Ibid.
119. Eligon, J. (2015, November 11). At university of Missouri, black students see a campus riven by race. *The New York Times.* Retrieved August 25, 2016, from www.nytimes.com/2015/11/12/us/university-of-missouri-protests.html?_r=1
120. *Tackling racial tensions on campus.* (2015, November 9). TimesVideo. Retrieved August 22, 2016, from www.nytimes.com/video/us/100000004027278/tackling-racial-tensions-on campus.html?action=click&contentCollection=us&module=embedded®ion=caption&pgtype=article
121. *Concerned student 1–9–5–0 presents list of demands to the University of Missouri.* (2015). Retrieved August 25, 2016, from http://bloximages.newyork1.vip.townnews.com/columbiatribune.com/content/tncms/assets/v3/editorial/3/45/345ad844-9f05-5479-9b64-e4b362b4e155/563fd24f5a949.pdf.pdf
122. Kolowich, S. (2015, November 20). Diversity training is in demand. Does it work? *The Chronicle of Higher Education.* Retrieved August 24, 2016, from http://chronicle.com/article/Diversity-Training-Is-in/234280
123. Basi, C. (2016). McDonald named interim Vice Chancellor for inclusion, diversity and equity: McDonald will begin serving in the dual appointment effective immediately. *University of Missouri.* Retrieved August 23, 2016, from https://nbsubscribe.missouri.edu/news-releases/2016/0613-mcdonald-named-interim-vice-chancellor-for-inclusion-diversity-and-equity/
124. Vishnani, A. (2016, August 24). MU holds first set of mandatory 'citizenship' trainings for new students. *The Maneater.* Retrieved August 26, 2016, from www.themaneater.com/stories/2016/8/24/mu-holds-first-set-mandatory-citizenship-trainings/
125. Bester, D. (2015). *A more progressive approach: Recognizing the role of implicit bias in institutional racism.* Retrieved August 22, 2016, from www.ncrp.org/publications/responsive-pubs/rp-archive/responsive-philanthropy-spring15/role-of-implicit-bias-in-institutional-racism. See also Feagin. (2006) *Systemic racism.*
126. *Diversity 101.* (2016). University of Missouri: Inclusion, Diversity & Equity. Retrieved August 24, 2016, from https://diversity.missouri.edu/education/diversity101.php
127. Ibid.
128. Fricke, P. (2016). *Mizzou will now require 'diversity intensive' courses for graduation.* Retrieved August 24, 2016, from www.campusreform.org/?ID=7412
129. Foley, H. (2016, September 20). Email communication.
130. *Report of the progress of the race relations committee to the faculty council.* Retrieved January 25, 2017, from http://facultycouncil.missouri.edu/issues/report-of-the-progress-of-the-race-relations-committee-to-the-faculty-council/RaceRelationsReport.pdf
131. Ibid.
132. *Faculty council committee on race relations: University of Missouri.* (n.d.). Retrieved August 25, 2016, from https://vimeopro.com/mizzouvideo/racerelations
133. Hudson, B. (2016, July 5). Personal communication.
134. Hudson, B. (2015, October 13). Diversity, inclusion training is right call for campus. *Columbia Daily Tribune.* Retrieved August 22, 2016, from www.columbiatribune.com/

opinion/oped/diversity-inclusion-training-is-right-call-for-campus/article_efbd88c0-5995-5385-ab6b-a01175561976.html

135. Pinegar, G. (2015, December 9). Race on campus. *VOX Magazine*. Retrieved January 25, 2017, from www.voxmagazine.com/news/people/race-on-campus/article_c4941089-0279-50aa-8fa8-1cb5702b3a19.html
136. Williams, M. R., and Hancock, J. (2016, November 2). The University of Missouri system meets its new president—Mun Choi. *The Kansas City Star*. Retrieved January 25, 2017, from www.kansascity.com/news/local/article112032492.html

2

LAYING THE GROUNDWORK FOR A DIVERSITY CULTURE SHIFT

> Diversity is not a fearful monster to be vanquished and buried under a rock, nor for that matter to be relegated to a swamp. Instead, its Hydra-like multiplicity ought to be embraced.
>
> —Sammy Basu, 2005, p. 23[1]

With the substantive insights gained from the "Mizzou" case study in mind, in this chapter we lay a conceptual foundation for the diversity change process by providing a definitional map of the key elements integral to a cultural shift. Following an exploration of the contested meaning of diversity in higher education, the chapter depicts the tensions that arise from contrasting structural features of the institutional environment including differing employment types and the "loose coupling" of decentralized units within a hierarchically-oriented campus landscape.[2] Our focus then shifts to the primary protagonists in a diversity culture shift and the ongoing evolution and challenges of the Chief Diversity Officer (CDO) role. From this vantage point, we highlight the emergence of pluralistic and culturally responsive leadership models and accent the increasing role of faculty governance as a countervailing force in championing diversity change. The chapter concludes with an overview of the critical components of organizational learning theory that undergird a diversity culture shift.

The Ongoing Counterpoint Between Diversity and Inclusion

Reaching consensus on the meaning of diversity given the multiple constituencies, divergent interests, and multifaceted dimensions of a campus environment is often an elusive proposition. In this sense, diversity on college campuses is not unlike the mythical nine-headed hydra that Heracles set out to conquer. Educators view diversity on a spectrum ranging from social conservatism through neutrality to social justice.[3] For example, a survey of 771 diversity officers, 63 percent of CDOs identified difficulties in arriving at a common definition of diversity on their campuses.[4] Yet the effort to eliminate different interpretations of what diversity means can damage the educational potential of diversity as a pluralistic and multifaceted construct.[5]

To unpack the complex issues relating to diversity and inclusion on college campuses, we need to interrogate the ways in which the meaning of diversity has evolved within the higher education context. While diversity is most commonly understood as difference and the valuing of differences, a second sense of diversity focuses on "the differences that differences make" and views the role of diversity leadership as reorganizing structures and processes that privilege some groups while disadvantaging others.[6] Or put in another way, diversity involves the "discourse of difference" and the multiple, complex, and sometimes contradictory ways that difference is understood, enunciated, represented, and addressed both within the context of institutional systems and the larger social context.[7]

As the "Report on Faculty Diversity and Inclusivity in FAS" (the Faculty of Arts and Sciences) at Yale University indicates, the opacity of the term "diversity" and the vagueness of the term "inclusivity" can be a way of avoiding discussion of difficult social topics such as racism, sexism, homophobia, etc.[8] The concept of inclusion transcends the attainment of compositional diversity and refers to the ways in which individuals are empowered to participate in the fabric of campus life. Specifically, inclusion represents the antithesis of asymmetrical power structures that suppress the voice of marginalized persons, limit self-determination, and circumscribe democratic participation.[9]

Viewed from this perspective, inclusion then involves concrete actions and practices that include participation in decision-making, having a voice, and distributive justice or access to resources on an equitable basis.[10] Within the workplace, inclusion refers to giving power to individuals who have been traditionally marginalized. It reflects the degree to which individuals are a part of core organizational processes and can contribute effectively to the organization.[11] Within the realm of higher education, inclusive practices require attention to staffing, policies, and mobilizing key constituencies such as tenured faculty, a process that can resemble "herding moody, drowsy lions."[12]

In a more neutral vein, the American Association of Colleges and Universities defines inclusion as active engagement with diversity in all aspects of college life. Such engagement fosters greater cognitive awareness and empathetic understanding of the interactions between individuals and systems or institutions.[13] On predominantly white campuses, however, an emphasis on engagement with diversity tends to focus more on the experiences of majority group members. By contrast, for members of nondominant groups, inclusion necessarily refers to how institutional practices, processes, and culture work together to create an environment that values collaborative input, ensures equity, and promotes mutual respect, recognition, and valuing of diverse perspectives.[14]

Norm Jones, Chief Diversity and Inclusion Officer (CD/IO) at Amherst College, emphasizes the ways in which diversity and inclusion differ, but have a reciprocal relationship. Jones, an African American male with an advanced degree in Organization Development, finds that organizations that attend to both diversity and inclusion simultaneously are the most nimble and adaptive:

> Part of the vision for communicating the relationship is to explain the ways in which those two things are not related. That is to say that when we think about diversity we think about quantifiable difference in composition as opposed to the work of inclusion which has to do much more with organizational culture and behavior. I like to think about the ways in which the work is both staged and also iterative; that is to say, it's difficult to deploy inclusive practices without having some level of compositional diversity.

> It also doesn't suggest because you deploy these practices that you shouldn't still continue to be attentive to compositional diversity. There's an inextricable relationship between those two things. I think that organizations that are attending to both realms simultaneously are the ones that are most nimble.

From a similar perspective, Brown University's Action Plan for Diversity (2016) emphasizes that "we must embrace both diversity and inclusion. It would be an empty victory to achieve one without the other."[15] Or as Harvard's first female president, Drew Gilpin Faust observes, "we must advance a culture of belonging—one in which every student finds and feels Harvard's opportunities fully available."[16] Inclusion then moves beyond the mere presence of differences to creating a sense of belonging and participation that is reflected in institutional actions, practices, and culture.

The Dangers of Neutralizing Diversity

Taking a look backward, the rise of the term "diversity" on college campuses has been shaped by Supreme Court decisions relating to affirmative action over the past four decades, beginning with the *Bakke v. University of California* case in 1978. In several landmark decisions, the high court has moved from support for remedial admissions efforts designed to address the historical underrepresentation of minoritized students to a singular rationale for diversity: i.e. the educational benefits of diversity. Yet ironically, the Court's diversity rationale suffers from inverted logic since it is primarily has focused on the educational gains of majority students on predominantly white campuses.[17]

The student movement has called attention to the need to re-center diversity goals and the "academic case" for diversity and inclusion on student learning outcomes. The "academic case" for diversity and inclusion is based on a significant body of empirical research that identifies the educational benefits of diversity and impact on student learning outcomes such as critical thinking, intellectual self-confidence, and cognitive skills as well as democracy outcomes such as pluralistic orientation, leadership skills, and civic engagement.[18] For example, a meta-analysis of 27 works with a combined sample of 175,950 undergraduates found a positive relationship between diversity and civic engagement in students' behavioral interactions, attitudes, and skills.[19] Given these findings, prioritizing of student learning will require the transformation of institutional culture and climate to create a welcoming environment and foster positive interactions across difference.

While diversity and inclusion arose historically in terms of the access of historically underrepresented students, in recent years, work related to diversity has evolved to focus on *all* salient identities within the prevailing political, economic and social context.[20] Viewed from this perspective, institutional diversity transformation is about changing the structures and milieu in which these identities are made salient.[21] Nonetheless, by using broad-based diversity terminology that includes all identities, attention can be diverted from discrimination and the reproduction of inequality through the use of non-accusatory and neutral language.[22] Yekim, an Afro-Latino male CDO at a private western university, underscores the risks of countering the prevailing trend to include everyone under the umbrella of inclusion. As he explains:

> Inclusion work is not about including everybody. And everybody assumes that to be inclusive means to include everybody. I have to be clear that inclusion work has historical, political, cultural and contextual underpinnings. That it's about righting wrongs. So inclusion is not about inclusion as much as it is about exclusion and how do we rectify exclusion. So who hasn't been at the table more so than get everybody at the table?
>
> When I think about inclusion in that way, it's always a risk. That's why when you hear so many CDOs . . . in the academic and scholarly spaces of presentation, everybody sort of tiptoes around the issue. Very few people really say what needs to be said. . . . most people dance around it. Part of it is the fear of how CDOs will be received on campus, and whether or not they will have their job when they come back, whether they will be terminated or whatever is the risk or consequence of fulfilling the role being asked.

When diversity includes everything, it can serve as a smoke screen that masks power differences between dominant and nondominant groups and fails to address historic barriers to inclusion.[23] An all-encompassing diversity concept fails to address how the "ideological apparatus" of oppression is replicated, reproduced, and transmitted through the prevailing systems, policies, and processes of higher education.[24] When defined as any difference, diversity "becomes everything and nothing, a signifier without a signified," and "a malleable object that can be deployed for various strategic purposes."[25] This emphasis overlooks the impact of centuries-long systemic forms of oppression and exclusion within U.S. institutions manifested in attitudes, behaviors, stereotypes, emotions, images, and practices.[26] Through a process of redirection from considerations of equity, diversity has been exploited as a convenient catch phrase and "rendered meaningless."[27] It can invoke a category of the other, a commodity, and term of convenience for corporate management that offers a sense of "racial innocence and absolution."[28]

Although "diversity" has often been conflated with race and ethnicity, it has more nuanced implications. It refers to the intersecting palette of social identities including the primary dimensions of gender, sexual orientation, gender identity, race, ethnicity, disability, and age that are protected by federal anti-discrimination law and executive orders, as well as the secondary and socially acquired characteristics such as socioeconomic background, religion, military status, educational background, marital status, and geographic location.

Over the past few decades, increasing scholarly attention has focused on intersectionality and how the multiple dimensions of identity intersect to affect an individual's experiences and social interactions. Yet not all dimensions have equal impact, since some identities are privileged and others are not. Women, minorities, and the poor, for example, are marked by their social location and are at least partially defined by it.[29] Dimensions of identity that reflect historical legacies of exclusion can compound to create multiple jeopardies such as when individuals are both African American and female or transgendered and disabled.[30] On college campuses, intersectionality frameworks can help educators understand the experiences of individuals at the margins of multiple groups as well as how the convergence of identities can contribute to inequality.[31]

Given our exploration of the implications of the meaning of diversity and inclusion, we now consider the issues arising from a highly differentiated campus environment characterized by both centralized and decentralized sources of authority and contrasting employment types and underlying values.

Surveying the Campus Landscape for Diversity

The landscape for diversity poses substantial challenges on college campuses due to structural contrasts: the co-existence of a centralized administrative hierarchy and decentralized units comprised of divisions, colleges, schools, and departments. This dichotomous organizational design has been described as a system of "loose coupling" in which the independent parts of a college or university do not act responsively together but instead often act independently.[32] Given the existence of conflicting power and authority structures, academic institutions are frequently characterized by internal disconnection, with struggles over identity, isolated decision-making, and a breakdown in shared meaning.[33]

The cultural divide between faculty and administrators represents a source of tension and even organizational dysfunction.[34] A lack of communication leads faculty to view the administration as a source of red tape and unnecessary bureaucratic constraints while administrators often see faculty as self-interested and unwilling to be accountable.[35] Faculty and administrators operate within different employment and authority structures: faculty authority derives from disciplinary expertise protected by tenure and academic freedom while administrators work within hierarchical conditions and respond to campuswide demands that are sometimes at variance with faculty values.[36] While tenured and tenure-track faculty work is characterized by functional independence and autonomy, nonacademic administrators are typically "at will" employees subject to the discretion of a singular supervisor. Frequent turnover among top administrators is symptomatic of the fragile and politicized nature of administrative employment.

Complicating the employment picture, a new workforce of part-time and full-time contingent faculty have come to represent nearly 70 percent of the faculty workforce.[37] Part-time faculty operate in an "at will mode" in a kind of "patronage" in which their employment is subject to the discretion of the department head and can be hired and fired informally without due process protections.[38] In essence, they also are in an at-will status similar to that of nonacademic administrators.

Despite the presence of loose and decentralized organizational coupling, the primary dimensions of power within the university or college still reside in a hierarchically based administrative structure governed by a systemwide and/or local boards of trustees. Over decades, the administrative hierarchy has remained structurally stable, reinforced by conventionalist leaders who have tended to exert influence through positional power.[39] Even when delegation of authority occurs, campus-based administrative officials including the president, provost, vice presidents or executive officers, and deans, hold considerable decision-making power.[40] Highly paid administrators control university budgets and have the most power, while students, particularly those from marginalized groups, have almost no power.[41] Given significant differences in employment status, interests, and values, administrators and faculty typically work in isolated silos and frequently only communicate among themselves.[42]

Other sources of authority in higher education include academic governance, departmental authority, and faculty or staff units.[43] As illustrated in the case studies at the University of Missouri at Columbia ("Mizzou") and the University of Tennessee at Knoxville (UTK), faculty senates can exert pressure on administration and provide a counterweight to external political forces through the power of shared governance. The principles of shared governance articulated in the American Association of University Professors' "Statement on Government of Colleges and Universities" adopted in 1966 identify the professoriate as having primary authority over curriculum in addition to faculty status, research, publication of research results, and the classroom.[44]

The increasing corporatization of the university is a further destabilizing trend that has led to an emphasis on efficiency emanating from the centralized leadership structure of administrators and nonacademic managers.[45] This business-oriented emphasis is in sharp contrast to the model of academic decision-making which emphasizes consensus and collaborative input. The focus on efficiency has often resulted from declining state budgets that have caused public institutions to seek entrepreneurial revenue sources in order to decrease costs. The constraints of the current budgetary environment, particularly in state institutions affected by legislative cuts, have diminished institutional resources for diversity efforts. Without the political will to safeguard these programs, diversity and inclusion programs are often the first to go.[46]

Given shrinking external resources, "academic capitalism" emphasizes the role of groups of institutional actors in deploying state resources to build channels to the new economy through interstitial organizations building channels to the new economy using state resources.[47] For such reasons, critics argue that campuses have focused more on business operations than the academic core.[48] In this increasingly fiscally focused environment, the role of the chief financial officer has assumed greater prominence and the position wields significantly more power in decision-making processes.[49] In models that bring the entire university budget under the provost such as at Princeton University and the University of Michigan, a sea change is underway in returning financial decision-making power to the academic side of the house. We also see the positive impact of appointing seasoned academics to presidential roles either as heads of campuses such as at Lehigh University or system heads as at the University of Missouri system.

In the face of conflicting external and internal political pressures, new leadership approaches are needed. In this context, organizational learning offers the potential to shift campus dynamics from reactive to proactive approaches that overcome internal/external dysfunction through a shared sense of purpose that transcends competing values and goals.[50]

Recasting Twenty-First-Century Diversity Leadership

Who are the primary protagonists in a diversity culture shift? Different perspectives in the literature suggest the power of grassroots efforts to drive change, while others emphasize the how executive administrative and academic leadership serves as the driver of change. While diversity change necessarily draws upon the consolidated input of faculty, staff, and students, presidential leadership is the *sine qua non* needed to instigate and sustain long-term change. A lack of senior leadership commitment has been identified as a major obstacle for institutions seeking to initiate pervasive diversity change.[51]

Research indicates, however, that there is no universal definition of leadership and that local definitions of leadership and culture impact how leadership is operationalized.[52] Furthermore, early models of leadership largely were derived from all male samples and these traditional perspectives differ from more relational and intrapersonal styles.[53] The concept of culturally responsive leadership refers to leadership practices that take race, ethnicity, gender, culture, and language into account as well as religion, disability, sexual orientation and social class.[54] To date, however, culturally responsive leadership research has, for the most part, centered on K-12 level scholarship on teaching and counseling.[55]

Scholars have called for more pluralistic leadership models that transcend traditional authoritarian modes and instead focus on empowerment and the sharing of power.[56] Such revolutionary leadership models are shaped by institutional context, promote learning and change, are process oriented, and emphasize collaboration.[57] These models deviate from traditional white, male, heterosexual perspectives by deemphasizing hierarchical relationships, acknowledging cultural differences, and reflecting awareness of one's positionality as a leader.[58] Rather than a values-neutral leadership paradigm, revolutionary leadership assumptions emphasize teamwork, partnership, collaboration, and interconnectedness with an acute awareness of the need for social justice and social change.[59] Revolutionary models also recognize the multiple, intersecting identities of leaders that occur within a particular context and the ways in which power dynamics have pervaded that context based on culture, history, and social structure.[60]

In higher education, leadership emanates from the president to the provost, executive officers, deans, department heads, and chairs. Leadership of the president and his or her Cabinet is critical in setting institutional direction on diversity, communicating a diversity vision, providing material support including budget, staffing, and needed resources, and ensuring accountability and follow-up. Marybeth Gasman, Professor of Higher Education at the University of Pennsylvania, identifies the necessary attributes of an organic diversity leadership strategy that engages vocal advocates at multiple levels and layers of the institution. Gasman, a white female faculty member, explains:

> The president has to lead these initiatives and then has to do so by working with the provost to get the deans to lead the initiatives and then get the deans to get certain faculty members who they know will buy in to lead the initiatives and then also work on the other side of the house getting certain staff members who they know are vocal and who are protected, and then working with students to lead the initiatives. I think you have to do this across the board simultaneously. But it has to come from the president, the president has to say it and has to show it. They have to regularly show that they really honestly believe that.

One of the conundrums of reliance on the institutional hierarchy for sustained diversity progress is the frequent turnover in top leadership positions. High-level administrative appointments such as the president, provost, and deans are subject to frequent turnover with the average appointment of a president as seven years.[61]

In championing a diversity agenda, presidents can be transactional in terms of following existing rules, developing exchanges based on existing norms, and negotiating through the use of power and rewards to influence change. Or they can exhibit transformational leadership by realigning culture through a new vision that addresses shared norms, values, and assumptions and promotes engagement and creativity in an environment that allows for self-actualization.[62] Transformational leaders can tie instrumental goals and objectives to their vision to motivate followers to higher levels of performance and overcome resistance.[63] More commonly, leaders combine both transactional and transformational styles in crafting a diversity agenda either in tandem or at different times.[64]

The president's own positionality can significantly influence his or her ability to shape institutional direction on diversity. For example, over half of the minority presidents, in a survey of 27 presidents, expressed concerns about being transformational diversity leaders due to the way white stakeholders could believe the diversity agenda was personal to them or simply reflecting self-interest, rather than the institution's interests.[65] Twenty-two of the presidents indicated that the most common barrier they faced in relation to diversity was resistance to a diversity agenda by members of dominant groups. The five presidents who did not identify this issue tended to come from single-serving institutions such as minority-serving schools and historically African American colleges and universities (HBCUs).[66]

Stereotypes about nondominant groups can lead to the perception that individuals from these groups do not have the wherewithal to hold leadership roles.[67] Berdahl and Min distinguish between descriptive stereotypes or generalized beliefs about what members of different racial groups are like and prescriptive stereotypes which, when violated, are likely to provoke social disapproval and backlash. Since, for example, East Asians in North America are often descriptively stereotyped as relatively competent, cold, and nondominant, when these individuals violate prescriptive stereotypes, assert their own viewpoints, or take charge, they are viewed as a competitive threat to valued resources that needs to be neutralized.[68]

Consider how Frank Wu describes his leadership experiences as Chancellor of the University of California Hastings College of Law and the importance of having a critical mass of minorities and women in the Cabinet:

> This is hard work, and if you think it's not going to be hard, it's going to be easy, you're going to be disillusioned and embittered and you're going to burn out. You have got to recognize that it's easy to get co-opted; it's much easier to make mistakes, then it is to do what's right. . . . I've spent most of my life. . . . being the only person in the room who looked like me. Even when I have been in responsible roles and leadership roles, I have sometimes been the only minority and I've been the one in charge, but it's easy when you enjoy privilege to not see the privilege you enjoy and the power you hold and to remember the importance of the issues you used to be passionate about. . . . I would say having a critical mass [is important].
>
> The best time for me was when my leadership team was predominantly female and predominantly black. . . . twenty-five years

> ago that was not true at UC Hastings or at any of the UC's. . . . You would not see a predominantly black, predominantly female leadership team headed by an Asian American. That doesn't happen often. We were not perfect, far from it . . . but we had a different understanding than a team that wasn't diverse. . . . it was the mixture, the ability to exchange ideas.

A critical factor missing from many leadership analyses is the instrumental role of the Board of Trustees in overseeing the university or college's strategic diversity progress. The president works closely with the Board of Trustees in translating board direction to institutional policy. Since in most public institutions the board is appointed by the governor with approval by the legislature or through statewide elections, board composition can reflect political pressures exerted by the prevailing political party.[69]

Boards of Trustees can serve as a significant check and balance to presidential power with the ability to insist upon the accountability of institutional leadership for diversity through review of policies, practices, and processes. Yet a study of board leadership indicates that boards often tend to rely on the president or chief diversity officer to provide periodic reports on diversity progress, without engaging in substantive review of ongoing institutional processes that impact equity.[70] As a report from the Project on Governance for a New Era indicates, trustees cannot simply play the role of cheerleaders or donors, but need to take a proactive role in exercising due diligence and benchmarking and reviewing the outcomes of the work of administrators and faculty.[71]

Without strategic and courageous leadership from the president and Board of Trustees that insists upon academic and administrative accountability and draws upon the collaborative input of tenured faculty and other key constituencies, diversity change efforts will result in piecemeal and fragmented activities rather than measurable outcomes. Committed leadership action is needed to undertake the difficult work of realigning institutional priorities, resources, and power structures. The reins of diversity leadership have in recent years focused on the Chief Diversity Officer as a member of the president's cabinet.

The Evolution of the Chief Diversity Officer Role

The evolution of the Chief Diversity Officer role over the past few decades has signaled the importance of diversity in terms of student access, institutional viability, and educational outcomes. Nonetheless, efforts to institutionalize diversity can create a stalemate in the context of larger, more stable academic structures juxtaposed with a smaller, nascent, and disempowered diversity structure.[72] Diversity offices are staffed by a small cohort of individuals with the Herculean task of providing programs to a university or college campus without the benefit of a supportive infrastructure. At the same time, seemingly institutionalized diversity programs can falter due to changes in institutional leadership and lose traction and momentum. From this perspective, realism demands that any apparently permanent, progressive change must be accompanied by a corresponding permanent institutional response.[73] Otherwise, momentary gains will only be a ripple in the water without staying power.

For many campuses, the expectation is that the chief diversity officer (CDO) will lead diversity change. Yet CDOs are often relegated to symbolic positions without the authority or leadership support to alter the status quo. Expected to do it all yet in a position in which they are subject to the perspectives of those above them including the president or chancellor and the board of trustees, CDOs function in a highly politicized, stratified political environment.[74] As Yekim, the Afro-Latino male CDO cited earlier explains:

> This role in itself is a risk. I think that anybody that takes on this role must go into the role with the understanding that they are at risk all the time. . . . the reality is that this work challenges power. It requires that a person . . . be an advocate of issues of equity and inclusion. That means not only by way of representation; but also by way of processes such as admissions decisions, tenure decisions, curricular content, how financial aid is awarded, among others. All those pieces that for me fall very well under consideration of diversity but most institutions don't really frame those processes as part of the diversity conversation.

With a title that seems to denote a certain level of power, CDOs may lack actual authority.[75] Their visibility and invisibility, high rank and limited resources, and titular recognition coupled with tokenism can lead to marginalization of their contributions.[76] For example, a case study of a Midwestern rural public state university reveals that the senior-level diversity position was established to offer the appearance of a commitment to diversity, while simultaneously distancing the CDO from hierarchical power and forcing the individual to persuade other institutional leaders of the benefits of diversity without sufficient funding or staffing.[77]

Ironically, the CDO role is the only senior leadership role in which the majority of individuals are diverse, with whites comprising only 12.3 percent of the incumbents in doctoral universities.[78] Within four-year degree granting institutions, 14 percent of all senior leadership positions are held by minorities, while African Americans hold 72.7 percent of Chief Diversity Officer positions.[79] This isolation suggests mere symbolic representation of the CDO in the upper echelons of a largely white higher education administration. In this capacity, the CDO can be marginalized as a tokenized diversity administrator functioning at the institutional periphery.[80] Given the underrepresentation of ethnic/racial minorities in high-level administrative positions, appointment of minorities to diversity-specific positions can have negative connotations, such as the following: 1) only minorities are interested or responsible for diversity work and 2) diversity leadership positions are the only executive positions that ethnic/racial minorities are competent to hold.[81]

Placed in this frustrating position, the CDO role can be extremely lonely. A study of five minority female CDOs, for example, found that several described their roles as "lonely," due to the potential isolation and marginalization.[82] One CDO reflected on what she called "the white privilege attitude at the highest level of the organization" including the president. She found her principal challenge to be the white privilege attitudes that she described as "very thick, at the very, top in the Cabinet, and by the leader of the Cabinet."[83]

Many CDOs serve as "at will" administrators, while relatively few have tenure status. When employed in "at will" status, diversity officers often lack the clout necessary to challenge the status quo. Even when a diversity officer is a faculty member with tenure, he or she may face substantial risks in developing diversity programs that are perceived as controversial or unwelcome. As Dawn, an African American diversity director in the College of Liberal Arts in a Midwestern public research university explains, the risks can be greater for minority women in this role when working with a predominantly white faculty who may be resistant to change:

> The major risk I guess is for me personally as an African American woman who has been doing diversity work but also in different types of positions . . ., and to come here and interact with faculty who have the same mentality as [where I worked] 25 years ago. One of the major risks that I find is I try not to internalize things so much, it can be very disheartening. . . . I think it would be a greater risk if I was not tenured: I'm not a full professor yet. I think about that about what I might say and whether they remember me when my portfolio for full professor might come up, whether they might be on the college committee or even within my department. That's a major issue that I constantly have to be reminded of.

Significant disparities in the way the CDO role is operationalized may signal the continuing reluctance of institutions of higher education to reframe diversity as a strategic imperative.[84] To date, the definition of CDO usually refers to the institution's highest ranking diversity administrator who reports to the president or provost or both and has a "boundary-spanning administrative role."[85] At the same time, institutions have begun to reconceptualize diversity leadership roles and create positions outside the executive office at the college or school level. Most commonly, such specialized positions report to the provost with a dotted line to the president and are focused on faculty diversity. Separate diversity leadership positions also have been established in health sciences and medical centers that are affiliated with university campuses. Nonetheless, unless diversity officer positions focused on students or faculty are connected to a campuswide effort, their efforts will not gather momentum.[86]

The redefinition and decentralization of the diversity officer role is consistent with the scope of our survey which includes both CDOs as well as diversity officers outside of central administration in key locations through a college or university. For example, Yale has appointed a deputy dean for diversity and faculty development in the college of arts and sciences. Princeton's vice provost for institutional equity and diversity serves as CDO, but a new position of dean for diversity and inclusion focused on students has recently been created. Stanford University does not have a CDO but locates diversity officers within each of its seven schools.[87]

The reluctance of diversity officers to participate in the survey conducted for this study likely derives from the lack of institutional support for their tenuous position within a predominantly white institutional hierarchy. This tension is particularly salient in state institutions that are subject to the scrutiny of conservative legislatures that have the ability to curtail funding for diversity initiatives and question institutional direction. In the wake of the election of Barack Obama, in 2008 and his reelection in 2012, many diversity officers were under assault personally, professionally, and academically.[88] Although initially CDOs imagined that social justice and equity programs would flourish under a minority president, instead the opposite has been true as attacks on public higher education have intensified. In this environment, equity/diversity workers were called upon to be "nice" when performing work that often involves witnessing cruelty and engaging in "hospitality work" in which all opinions and experiences are assumed to be valid.[89] The advent of Republican candidate Donald Trump to the presidency has placed the CDO role in an even more tenuous and uncertain position.

Faculty Governance and Diversity Change

The role of faculty governance in diversity leadership has become increasingly important in an era of external political instability, particularly in public higher education. In the past, faculty governance typically centered on academic decision-making in curricular-related matters and often has not taken front and center stage on issues of diversity. As seen in the case studies at both the University of Missouri and the University of Tennessee at Knoxville, faculty senate leadership has been instrumental in moving forward on a proactive diversity agenda, especially in the face of regressive legislative pushback. In addition, grassroots faculty leaders can serve as change agents for social justice even when they are not being associated with positions of authority such as chairs of the faculty senate or department heads.[90] Yet even faculty senates can become mired in bureaucratic practices. As Marybeth Gasman observes:

> Faculty senates are usually also filled with people who are really bureaucratic or will say nothing. They've learned how to give a speech. I don't think that does anything. Some of the best efforts are when you have people organically come together and rise up and make change. I have seen people from the faculty senate say the same stuff over and over and they issue statements. . . . I participate in faculty governance wholeheartedly, but when you lose touch with what's going on in reality and then end up talking in policy speak, it's just not going to do anything. You have got to remember to get down on the ground and actually work with people. I just don't see that happening.

While the role of tenured faculty is critical in championing diversity, many tenured faculty will nonetheless be reluctant to speak up. The risks of faculty activism are real even for tenured faculty, due to the potential for collegiality to be questioned and the threat of jeopardizing tenure, promotion, or even contract renewal.[91] Faculty also must juggle multiple priorities including expanding teaching and service roles, academic capitalism involving grants and external funding, and pressures of tenure and promotion.[92] These pressures can preclude faculty willingness to take on more public stances on diversity issues. Yet at the same time, Gasman notes the fears that keep even tenured faculty from speaking up:

> One of the reasons why I am fiercely protective of the tenure system is that I think tenure is one of the things that allows faculty to fight back against these systems that oppress people. It's just whether or not they will do it. I was in a room the other day where I was talking to about 150 tenured faculty members; after I finished talking, a whole bunch of them asked me, "How did you become so brave?" I basically looked at them and said, "What are you afraid of? What on earth are you afraid of? What is going to happen to you?" I think most people just don't have it in them to push back.

Yet even when administrative and academic leaders seek to implement diversity change, they can face perilous undercurrents within the prevailing culture of an institution. As one research study reveals, the challenge for presidents is to find the right pace of cultural change so as not to get ahead too rapidly or provide so little impetus that the campus becomes complacent.[93]

Organizational Learning as a Catalyst for Diversity Transformation

With the contrasting dimensions of the campus landscape in mind, the aspirational goal of diversity organizational learning is to build the capacity of all members of the campus community to work collaboratively across difference. In this effort, campus leadership must attend to how different identities are embedded within the culture and practices of the institution.[94] Kezar refers to the greater simplicity of communication when campuses were more homogenous, indicating that "when the majority of the faculty and staff came from similar economic and social backgrounds, there was a greater alignment of interests and ease in understanding each other."[95] Yet not surprisingly, in the past and even in the present, the burden of understanding others has generally been placed on individuals from nondominant groups. A study of the websites of 80 public institutions in 2009, for example, found that even in institutions that expressed a commitment to diversity, diversity was seen often in terms of the integration of "others" rather than in relation to the transformation of the entire campus community.[96]

Despite the proliferation of management research on organizational learning, the concept itself is not well understood or even widely used or recognized within higher education. The concept of organizational learning arose in the 1950s among cognitive psychologists such as Herbert Simon and has permeated the management literature for several decades.[97] Cautioning against the danger of reifying organizations as "learning" or "knowing" something, Simon described learning as occurring through the learning of the organization's members or through knowledge brought by new members. He identified a major topic in organizational learning as understanding the mechanisms by which an organization can deviate from its culture.[98]

The research of Chris Argyris and Donald Schon dominated the field of organizational learning in the 1990s and is particularly germane to how organizations learn about diversity.[99] Based on research findings involving 6,000 individuals, Argyris describes two master designs or "master programs" that encompass how humans deal with others: 1) the espoused theory individuals use to justify or describe their actions and 2) the actual theory-in-use. When an issue causes embarrassment or threat, a systematic discrepancy has been found between the espoused theory and the theory-in-use as well as a systematic lack of awareness of the discrepancy at the time it is produced.[100] While espoused theories vary, the theories in use are remarkably consistent. Defensive routines that come into play in a threatening or embarrassing situation include the following: the need to be in control, striving to win and not lose, suppression of negative feelings, and the effort to be as rational as possible.[101] Similarly, on an organizational level, organizations will not be motivated to learn when information threatens the collective self-concept or is anxiety producing. To overcome these ego defenses, organizations must actively promote dialogue on future identity and embrace the concept of a learning organization.[102] When faced with threatening or anxiety-producing situations, organizations will follow three fundamental rules: 1) bypass the error or issue and act as if it were not a concern; 2) make the bypass itself "undiscussable;" and 3) make the undiscussability undiscussable.[103]

Two main types of learning apply to diversity strategy. Single-loop or first-order learning is incremental, instrumental, and adaptive and corrects errors through changes to routine behavior without addressing the values of the underlying theory-in-use.[104] Such learning is primarily tactical and superficial, since it leads to "projectitis" or an additive process of creating one diversity initiative after another without engaging the deeper issues.[105] By contrast, double-loop learning or second-order learning corrects errors or mismatches by addressing underlying values of the theory-in-use as well as strategies and assumptions.[106] For example, if an institution of higher education seeks to build a more welcoming climate, the value of inclusiveness may be an espoused theory in its mission statement but not actualized as a theory-in-use.[107] To attain a more welcoming climate, strategies and assumptions that translate espoused theories into action must be adopted.

Building on the frameworks of Argyris and others, management theorists at the University of Michigan Business school identify *organizational learning capability* as the capacity to "generate and generalize ideas with impact (*change*) across multiple, internal organizational boundaries (*change*) through specific management initiatives and practices (*capability*)."[108] Reduced to a formula, organizational "learning capability = ability to (generate × generalize ideas with impact)."[109] This process necessarily involves correcting "learning disabilities" within an organization.[110]

Based on the wealth of management literature on organizational learning, ten key characteristics have particular relevance to the higher education environment. These defining characteristics provide criteria for gauging the effectiveness of diversity organizational learning efforts. The criteria raise important questions about how organizational learning is embedded within institutional practices, draws on collaborative perspectives, and engages the campus community as a whole.

Organizational learning:

1. Focuses on how and under what conditions the institution learns and changes.[111]
2. Generates and generalizes ideas with impact across multiple organizational units.[112]
3. Addresses the processes by which organizations learn.[113]
4. Engages the institution's core work and is linked to organizational identity.[114]
5. Employs common structures and involves cross-institutional stakeholders to reflect on both process and results.[115]
6. Enables a critical mass of individuals to operate in new ways resulting in the creation of new habits and norms.[116]
7. Addresses ways of changing organizational culture and improving organizational actions.[117]

8. Can occur in "pockets" since organizations generally do not learn as a whole.[118]
9. Is designed to enhance effective action, not simply thought or insights.[119]
10. Requires conditions of psychological safety to allow interpersonal risk-taking.[120]

A prominent characteristic of diversity organizational learning involves reframing and deframing prevailing exclusionary frameworks including the white racial frame, male sexist frame, and heterosexist and ableist frames.[121] The case studies describing the work of the Faculty Race Relations Committee at the University of Missouri at Columbia as well as that of the Office for Diversity and Inclusion at the University of Maryland offer specific steps that institutions can undertake to implement such deframing and reframing.

Organizational learning is frequently confused with organization development (OD), an allied concept which refers to systematic interventions designed to foster organizational effectiveness over a sustained time period. Based on the foundational work of Kurt Lewin and his collaborators at the National Training Labs in Bethel, Maine, the field of organization development emphasizes the relationship between research, training, and action.[122] Lewin articulated the concept of action research with three driving principles: 1) change requires action; 2) action requires analysis of the current situation, identifying alternatives, and choosing the most appropriate one; 3) individuals must believe that change is necessary and need to reflect on the totality of the situation to gain new insights into the change process.[123] Lewin's three-step change model of unfreeze-change-refreeze recognizes the need to destabilize organizational equilibrium in Stage 1 due to the threat that change poses to the status quo. During Stage 2, the motivation for change occurs as new responses develop to new information, while in Stage 3 the change is stabilized and integrated.[124]

Organizational learning and organization development have clear points of overlap, but also a differential focus. Both address changing culture and the improvement of institutional actions. However, organizational learning is geared specifically to the process of learning in terms of how it occurs in terms of approaches, different constituencies, and level of engagement. From a process perspective, diversity learning may resemble a "patchwork quilt" rather than a standard process applied uniformly across an organization.[125] Due to a rigid, top-down bureaucratic structure, social learning can be an espoused goal rather than a practical reality and despite their professional expertise, faculty are often not called upon to create knowledge for their own institutions.[126] Or diversity learning can be approached on a piecemeal basis in which diversity projects are substituted for diversity progress.[127]

To conclude the chapter, we introduce a case study of the University of Tennessee at Knoxville (UTK) that provides a startling example of the intrusion of conservative legislators in the area of diversity within public institutions. It also illustrates the deleterious impact of such intrusion on university administration and the courageous leadership role of the faculty senate that emerged as a countervailing force. Following the case study, in the next chapter we introduce a conceptual framework for systematic diversity organizational learning and illustrate the evolution of the change process with specific examples.

Case Study II
Crossing the Line: Legislative Threats to Diversity Programs at the University of Tennessee at Knoxville (UTK)

This case study examines the pressures exerted by the Republican-dominated Tennessee state legislature on the University of Tennessee at Knoxville (UTK), the flagship campus of the University of Tennessee system, regarding the content of diversity education and programming. It points out the dramatic contradiction between the university's strategic plan with its stated diversity goals and a bill passed by both houses of the state legislature that defunded UTK's Office for Diversity and Inclusion for fiscal year 2016–2017. The state law placed restrictions on using state funds to promote the use of gender-neutral pronouns, support or inhibit celebration of religious holidays, and fund or support Sex Week. The law is destined to sunset after 2016–2017 unless legislators vote for an extension.

The unprecedented legislative action of modifying a line item within a university's budget followed threats for even more sweeping budgetary cuts to diversity programs. In addition, legislators called for the firing of the 24-member University of Tennessee Board of Trustees, Chancellor Jimmy Cheek, leader of the UTK campus, and the highest-ranking African American official at UTK, Ricky Hall, Vice Chancellor for Diversity and Inclusion. The Tennessee state legislature funds approximately 23 percent of the UTK's budget and over 30 percent of the UT system's budget and has used this leverage as a method of control over university programs.

Geographic Location and Historical Legacy

UTK is located in East Tennessee (Knox County), a region considerably less diverse than other urban areas including Memphis in western Tennessee (Shelby County) and Nashville in central Tennessee (Davidson County). Whereas in 2015 the overall African American population of Tennessee was 17.1 percent and the Hispanic population 5.2 percent, the population of Knox County was 9.1 percent African American and 4 percent Hispanic. By contrast, Davidson County's population was 28 percent African American and 10 percent Hispanic while Shelby County's population was 53.5 percent African American and 6.1 percent Hispanic.[128]

UTK only began to gradually desegregate its enrollment in 1951 following a district court order that ruled in favor of African American applicants to its graduate and professional programs. The first African American undergraduates were not registered until January 1961 and included Theotis Robinson, Jr. who later became Vice Chancellor of Equity and Diversity of the UT system.[129] Robinson had been denied admission a year earlier based on UT policy not to admit African American undergraduate students. The Board of Trustees changed the policy under threat of a lawsuit by Robinson and his parents.[130]

Leadership and Governance at UTK

The complex, multi-layered structure of governance of higher education in Tennessee directly impacts the ways in which UTK operates. About half of the states consolidate public higher education institutions under one or more super-governing boards.[131] The Tennessee Higher Education Commission (THEC) was created by the Tennessee General Assembly and coordinates two systems of higher education: the University of Tennessee institutions that are governed by the Board of Trustees as well as universities and colleges that are under the purview of the Tennessee Board of Regents. The Commission reports to the governor, Bill Haslam, a Republican, and is accountable to the legislature, As such, it holds considerable power with the ability to review and approve new academic programs, new academic units, and new instructional locations.[132] The governor also heads the Board of Trustees for the University of Tennessee system.

A system president oversees the work of the University of Tennessee institutions. In addition to the flagship UTK campus, the University of Tennessee system includes campuses at Chattanooga, and Martin, a health sciences campus in Memphis, as well as a Space Institute at Tullahoma and two other statewide institutes. UTK is a public doctoral research university with an enrollment of approximately 28,000 students. Twenty-four percent of students identify as non-white, including 3.5 percent Hispanic, 3.1 percent Asian or Pacific Islander, 6.8 percent African American, and 2.7 percent two or more races. In addition, of the 1,526 full-time tenure-track or tenured faculty, 10.1 percent are Asian/Pacific Islander, 4.1 percent are African American, 3.8 percent are Hispanic, and 80.9 percent are white.[133]

Over the last 16 years, both the Tennessee university system and the UTK campus have experienced frequent administrative turnover with eight different system presidents and four different campus chancellors. Three system presidents were fired by the Board of Trustees and a highly respected campus chancellor, Loren Crabtree, was fired by the system president who was then, in turn, fired by the Board.

In 2012, UTK Chancellor Jimmy Cheek, a white male appointed in 2009, created a new position of Vice Chancellor for Diversity and Inclusion and later hired Rickey Hall as the inaugural Chief Diversity Officer. Although the new diversity office initially had no staff, it was reorganized to include the Pride Center, the Office of Equity and Diversity, the Office of Multicultural Student Life, and several diversity-related commissions.

In 2010, then-governor Phil Bredesen challenged UTK to set priorities leading toward the goal of becoming a top-25 research university.[134] The University's strategic plan, VOL Vision 2015, was developed in response to this challenge and later updated to VOL Vision 2020 in 2016. VOL refers to a three-part vision for value creation of knowledge for a global, multicultural world; original ideas that discover new knowledge and generate new ideas; and leadership through graduates with the capacity to be ethical and capable leaders.[135] Diversity and inclusion is one of the six priorities identified in the strategic plan and includes the goals of improving campus climate, recruiting and retaining faculty, staff, and students from diverse backgrounds, and preparing students for careers in a diverse, global society.[136] In 2014 UTK also developed a diversity impact plan in response to accreditation requirements of the Southern Association of Colleges and Schools (SACS) as well as a Quality Enhancement Plan (QEP) focused on engaged learning.

Beauvais Lyons, President-Elect of the Faculty Senate and a Chancellor's Professor in the School of Art, describes the objectives articulated in these documents as focusing on student preparation for citizenship in a global society:

> Our initial Quality Enhancement Plan for SACS accreditation focused on "Ready for the World," in response to the profile of our students who principally come from the state of Tennessee; and the effort was in order to prepare them to be part of a global workforce, and for life after college, to gain an awareness of the world outside of Tennessee.

Clashes With the State Legislature

Conflict with the state legislators over UTK's diversity programs arose over three issues: Sex Week, informational web postings on gender-neutral pronouns, and inclusive holiday celebrations. The first issue arose in 2013 when Sex Week was developed by students at UTK following the example of other universities that had hosted such programs in order to give students better access to information about sexuality. More than 4,000 students attended the first Sex Week program. In response, Tennessee legislators threatened to punish the campus for using student fee funds to support the program by preventing the University of Tennessee and other public institutions of higher education from using institutional monies to pay guest speakers.[137] The UT system subsequently issued a policy directing the Knoxville and Chattanooga campuses to require students to expressly authorize an opt-in fee for student-fee-funded programs. A resolution by the Tennessee State House specifically condemned the students by name who had organized Sex Week, calling it "an atrocious event."[138] Although the resolution was not formally adopted, over a period of several years, the legislature slowly and inexorably moved toward eliminating state promotion and support of the event.

The second event that led to further outcry by the Republican-dominated legislature was a memo posted on the Diversity and Inclusion Office's website on August 25, 2015 by Donna Braquet, director of the Pride Center. The memo sought to educate the campus community about gender-neutral pronouns and the preferences of transgender and gender nonconforming individuals for the pronouns "they" or "ze" rather than "he" or "she."[139] The Pride Center provides resources for LGBTQIA (lesbian, gay, bisexual, transgender, queer, intersex, and asexual) and ally students, faculty, and staff at UTK.[140]

President Joe DiPietro, president of the University of Tennessee System and a veterinarian by training, stated that the controversy was unlike anything he had ever seen.[141] He indicated that while the social issues were appropriate ones for discussion, "it was not appropriate to do so in a manner that suggests it is the expectation that all on campus embrace these practices."[142] On September 4, Dr. DiPietro announced that in consultation with Jimmy Cheek, Chancellor of the UTK campus, the memo had been taken down from the website. He further explained that Chancellor Cheek would advise the vice chancellors "not to publish any campuswide practice or policy without his approval after review with the cabinet."[143] The coordinated actions of DiPietro and Cheek reflected the administration's concern regarding the potential of the legislators to defund university programs.

The third event arose in December 2015 when the UTK Office for Diversity and Inclusion posted a blog that recommended avoiding religious symbols at holiday work parties. The blog provided a list of suggestions for inclusive celebrations that are respectful of students, colleagues, and the university and recommended that employees ensure that "your holiday party is not a Christmas party in disguise."[144] In a Fox News interview, Representative John J. Duncan, Sr., a Republican, warned against the "political correctness" involved in these two actions and stated that the liberals in the United States are the most intolerant in the country.[145]

At first, the Office for Diversity and Inclusion posted a note on the website, stating that the blog was only advice and not university policy.[146] On December 3, calls for the resignation of Chancellor Jimmy Cheek by members of the legislature followed. Some lawmakers called for Vice Chancellor for Diversity and Inclusion, Rickey Hall, to resign as well. Petitions to support Cheek and Hall gathered more than 3,000 faculty, staff, and student signatures.

On December 8, Chancellor Cheek issued a press release referring to future efforts to "prevent further poorly worded communications that deter from making progress on the [diversity and inclusion] goals."[147] Backing down on the holiday controversy, Cheek announced that Vice Chancellor Ricky Hall had provided leadership on improving the campus culture and environment, but had been "counseled." His apology stated, "We are sorry that we did not communicate very well. We've learned a lesson from this." Oversight of the Office for Diversity and Inclusion's website was reassigned to the Vice Chancellor for Communications.[148] System president DiPietro indicated that future posts would be reviewed by Chancellor Cheek and his cabinet.

Since the web postings were informational and did not represent university policy, Matthew, a white male faculty member, believes a precedent was established by the UTK administration's posture:

> There were several key decisions that I think most of the faculty, not all, think were mistakes. . . . the university initially put out a statement pointing out that it wasn't policy, just a recommendation. . . . but then the Chancellor in consultation with the president decided to take down that post, to delete it basically. . . . part of his statement said that Rickey Hall would be 'counseled' and a lot of people thought that word was really demeaning, and people still talk about it.
>
> Now maybe it was just a momentary lack of words. But it was an extremely poor choice of words in the opinion of many faculty. So I think many faculty believe that that whole situation was just mishandled. Many faculty believe that the post should not have been taken down, that maybe some clarification should have been added. Although perhaps technically it wasn't a violation of academic freedom, it came awfully close to it. And that in some sense because we gave in on that point, that was the beginning of the end.
>
> The way that was handled, by suggesting that Rickey Hall needed some sort of counseling, and then basically backing down . . . , I think that was a real turning point. . . . and that

> somehow this then has to be put under the Vice Chancellor of Communications. . . . that was the whole ball of wax; the whole "solution" to the problem I think faculty thought it was wrong from one end to the other. . . . it was a solution that really satisfied nobody. . . . many of the faculty say that set a precedent for this low level interference by the legislature in the business of the university.

On December 8, Representative Micah Van Huss, a Republican, warned that he was sponsoring legislation to defund UT Knoxville's diversity office. In his view, the "so-called Office of Diversity" was "not celebrating diversity, they are wiping it out. It is the office of Political Correctness."[149] Stating that the memo on gender-neutral pronouns "in no way reflects the values of my constituents," Van Huss expressed concern about the UT system's diversity funding of $5.5 million from its $2.1 billion budget for all its campuses including Chattanooga, Martin, Memphis, and Knoxville.[150]

In a television broadcast on December 13, 2015, State Representative Martin Daniel, Republican, indicated it would take legislative action such as defunding the diversity department, limiting its spending, instructing the diversity office to only be involved in recruitment and discrimination education, and/or firing the Board of Trustees. He described the diversity funding as "feel good money" to make the students feel better and added:[151]

> I have never seen a supervisional body that's more disengaged from what's going on than the UT Board of Trustees. Everything that goes according to what the administration wants. The diversity program has been set up for one reason, that is to protect the backs of the administrators, to protect them from minority group, special interest group criticism. I am telling you that the legislature is looking very seriously at reconfiguring the Board of Trustees, unless they set up and be more engaged, listen to what the people are saying. We don't appreciate this kind of thing coming out of our public university, our flagship university. They need to listen to the people, listen to us, or changes will be made.

Chiming in on the television debate, a political analyst, George Korda, indicated that the protests around the country were over "moronic issues" and said the Office of Diversity was a "means by which to cut off dissent, to cut off debate."[152]

Capping these three controversial events at UTK, in January 2016, ten Republican representatives and state senators led by state representative Eddie Smith wrote to Tennessee's Speaker of the House, Beth Harwell, and the Lieutenant Governor, Ron Ramsey, both Republicans, requesting a joint committee be formed to investigate the diversity offices at the state's public universities and colleges. According to Smith, "the university needs to reflect the values of Tennessee."[153] The letter stated that the legislators wish to review which people are involved in diversity, how money is spent, as well as the "actual productivity and efficiency of persons engaged in diversity activities."[154]

Finally, during the week of April 19, 2016, both the House of Representatives and the Senate voted to remove the $436,000 state appropriation for the Diversity and Inclusion office at UTK for the 2016–2017 year. This funding constituted 0.02 percent of the University of Tennessee's budget and included $260,000 for salaries for 3.75 FTE and 0.25 FTE for the Director of the Pride Center.[155] Instead, the allocation was to be directed toward the funding of minority scholarships in engineering.[156] As Matthew, the white male faculty member cited earlier, speculated:

> I think there is a strong element of homophobia, there's probably an element of racism there too. In terms of minority scholarships: well, engineering is a worthwhile thing to study; not the humanities certainly! We don't want to be called racists. So we'll say minority scholarships for engineering. I really don't know that it was a carefully considered decision.

The bill also banned the expenditure of state funds "to promote the use of gender-neutral pronouns, to promote or inhibit the celebration of religious holidays, or to fund or support Sex Week."[157] An earlier version of the bill would have used $100,000 of the funds to place decals on law enforcement vehicles that read "In God We Trust."[158]

In response to the legislature's efforts, hundreds of University of Tennessee students marched in protest with some demonstrators lying down in a walkway and later forming a circle in front of a student residential building. As the protest continued, Confederate flags were hung outside two dorm windows. George Habeib, a 19-year-old printmaking student, described the hanging of the Confederate flags as "trying to instill fear in us" and added, "it further proves our point of why this [sort of protest] is needed."[159]

Nonetheless, a student coalition, UT Diversity Matters, presented a list of demands to the administration including the need for mandatory inclusivity training and LGBTQ+ for all incoming students, faculty, and staff. The students demanded that Inclusivity training analyze oppression, privilege, and intersectionality as well as systematic, interpersonal, and internalized oppression.[160] Mandatory LGBTQ+ and inclusivity training developed by a committee should discuss historical discrimination against LGBTQ+ individuals, current issues including violence, harassment, and suicidality, and the experiences of specific groups included under the LGBTQ+ umbrella.[161]

On May 20, 2016 Governor Bill Haslam permitted the controversial bill originally known as HB 2248 to become law as Public Chapter 1066 without his signature, in effect abstaining from taking a position. The diversity office subsequently closed. Rickey Hall took a position at the University of Washington and subunits of his office such as the Office of Multicultural Student Life were reassigned to other Vice Chancellors.[162] The administration cut back on the Pride Center which was focused on facilitating Safe Zone training and providing a safe and inclusive environment for the LGBTQ+ community.[163]

Responding to legislative incursions on the Knoxville campus, system president DiPietro pointed out that while the campuses of UT have become more diverse, they have not necessarily become more inclusive.[164] He noted incidents of bananas thrown at African American students who had come for a visit and cotton balls thrown in front of African American cultural centers and explained why the university needed to address multicultural inclusiveness as an integral part of the culture:[165]

> And that says to me that we have a lot of work to do with our own house, and its faculty, staff and students, to make them more competent around multicultural inclusiveness as part of the fabric of what we do. I think where we've had trouble is in making inclusion and civility part of the fabric of our institutions. And that's what these offices of inclusion and diversity are meant to do.

DiPietro acknowledged that while the legislature had a right to hold him accountable, the role of the governing board is to provide direction about programs and direction of the university system.[166]

On June 21, 2016, President DiPietro announced that Chancellor Jimmy Cheek was resigning and returning to the faculty after eight years of leading UTK. Challenges that clearly may have led to Chancellor Cheek's resignation included his efforts to seek middle ground on the diversity controversy and a Title IX lawsuit settled for $2.48 million over the handling of sexual assault complaints filed by eight women.[167]

Taking a Proactive Stance on Diversity and Inclusion

The new law defunding the Office for Diversity and Inclusion passed by the state legislature poses significant dilemmas for UTK in its efforts to continue diversity programming, maintain academic freedom, meet accreditation objectives, and attain its aspirational goal of becoming one of the top 25 universities. Bruce MacLennan, 2015–2016 President of the Faculty Senate and associate professor of electrical engineering and computer science, describes the impact of the legislation on recruitment of faculty, staff, and administrators:

> Frankly, one of our goals is to become a top 25 university. We are trying to give a quality twenty-first century education. We are fighting even to try to keep up with other universities, especially on some of these issues. While we have made progress on our diversity and inclusion efforts, in student recruitment especially, and some on faculty, we're still nowhere near where we want to be. We have heard from faculty and administrators that recruits have said, "Well, I was thinking about coming to UTK, but I have changed my mind because of all the stuff that's going on." It's hurting us in all sorts of seen and unseen ways. That's very discouraging.

Notably, the UTK Faculty Senate has assumed a significant leadership role in diversity and inclusion in response to the state legislature's actions. Professor Beauvais Lyons, who was previously Faculty Senate President in 2003–2004, is also chairing a newly formed Faculty Senate Task Force on Diversity and Inclusion. Lyons' return to this role was in response to the need to have experienced faculty leaders advocating for the academic mission of UTK.

The UTK Faculty Senate has authored several key resolutions that were, in turn, forwarded to the Board of Trustees and the governor. These resolutions address the undue influence of the legislature on the University of Tennessee as well as the importance of diversity and inclusion.[168] Bonnie Ownley, Professor/Graduate Director of Plant Pathology who serves as incoming President of the Faculty Senate and former Co-chair of the Senate Committee on Diversity, emphasizes the need to be proactive rather than reactive and the importance of faculty partnership with administration. Noting the trend nationally for legislators to attack public higher education, Ownley expresses concern about the potential for deeper incursions into campus research agendas and academic programs:

> At some point we have to be proactive instead of reactive; at some point it seems like every time the legislature has a victory it pushes them to do more . . . that is very concerning to

> me. Maybe it does depend on what happens in November [in the elections]; it's almost overwhelming. It's just an attack on higher education across the country.
>
> What worries me next is if it's diversity today, what's it going to be tomorrow? Are they going to start trying to edit and manipulate the content that we have in our classrooms? And we know there are three topics which are near and dear to their hearts: one is climate change, the other is stem cell research; and the third is microaggression research. Those are things they would not like to see in the classroom. What's going to happen? How far are we going to let them go with this? And it's not just Tennessee, it's across the country. I think we as faculty really have to work with our administration as best we can. It doesn't do you much good to say we're going to have vote of no confidence when nobody cares.

In a similar vein, Faculty Senate President-Elect Beauvais Lyons addresses the need to be proactive "as the only way to move forward" rather than simply doing damage control. As a key aspect of this strategy, Lyons recommends that faculty and students work with the Education Committee of the state legislature to communicate the importance of diversity and inclusion. He further underscores how campus diversity programs benefit the recruitment and retention of diverse students:

> When you have the parents of gay or lesbian or bisexual or transgender freshmen come to campus concerned that there is a safe place for their son or daughter to have a supportive community, we recognize that the Pride Center is important to our own retention goals, just as the Frieson Black Cultural Center is to retaining African American students, just as our office of student disability services is to be helping a student with, for example, physical disabilities.

Lyons also points out that faculty can advance diversity goals separate from campus-level funding, which is limited by state law. For this reason, he indicates that the Faculty Senate plays a critical role in terms of reminding faculty of the principle of academic freedom and its protections that are essential to the existence of a free university. As he explains:

> We know that in our interpretation of Public Chapter 1066, that with regard to the university now being required by law not to promote the use of gender-neutral pronouns that the campus administrators essentially have to comply with the law as officers of the university, but the faculty members themselves because of our rights to academic freedom, can include information on our syllabi about gender neutral pronouns, can speak openly from our positions as individuals who reflect our disciplines or our own conscience. This is what makes for a free university. To a great extent the Faculty Senate can help to advance the diversity goals in ways that campus administrators may be limited by law from doing.

Professor Louis Gross, Chair of the Faculty Senate Budget Committee and professor of ecology and evolutionary biology, also stresses the critical role of the university in protecting academic freedom:

> There certainly is a feeling here that a university has to stand up to any infringement of individual freedoms and academic freedoms. It is very clear in the faculty handbook that the faculty have control over the curriculum and the academic aspects of the university.

Confirming this perspective, Professor Joan Heminway, former faculty senate president and professor of law, describes the effect of years of compromise with legislative critics on the right of free speech as well as on educational communication on campus:

> We are in a position where we just keep rolling and rolling, and so they ask for a mile, and we give them maybe more than an inch, maybe half a mile. And for the next year they ask for two miles, and we have to give them a mile. We are in this place now

> where we are falling behind in protecting, in my mind, the very important rights of free speech and educational communication on our campus. . . . I guess the way I would characterize it is that is has been political compromise for five years and that has led us down a perilous path.

Other efforts by the Faculty Senate include a forum held on June 17, 2016 at which representatives from the Pride Center discussed their work including Safe Zone training. John Zomchick, Vice Provost of Faculty Affairs, was the only administrator present to respond to a crowd demanding answers and expressing sadness and hostility over the university's implementation of the new law and what some perceived as a "wavering commitment to diversity."[169] Lyons intervened in the meeting to note that all were on the same team and reminding the audience not to blame the administration for the legislators' actions. Donna Braquet, former Pride Center director, said if UTK administration was committed to the ongoing work of the Center, should have consulted her regarding funding options but has not contacted her.[170]

Diversity programs that have continued at UTK including training on compliance issues as well as STRIDE training ("Strategies and Tactics for Recruiting to Improve Diversity and Excellence") for faculty search committees. The STRIDE training is based on work developed as part of the ADVANCE program funded by the National Science Foundation at the University of Michigan designed to increase the participation and advancement of women in Science and Engineering fields and focuses on academic research on implicit bias and diversity.[171]

Next Steps

Given the events that have occurred on the UTK campus, Bruce MacLennan believes that student activism for diversity is not likely to taper off, but will to some extent depend on the national climate as well as the direction of the administration. As he explains:

> I doubt it is going to taper off because the mood of the country does not seem to be tapering off; and they're going to take some cues from what's going on nationally. . . . if the students . . . see that the administration is making a wholehearted effort within the limits of the law and has demonstrated a strong commitment to diversity and inclusion, I would still expect to see increased student unhappiness just because of the national climate and because the issues have grown. That is perhaps the good outcome. The bad outcome would be that the administration seems to be retreating or embarrassed about its diversity and inclusion efforts, then I think students will be much more angry and will be pushing back even harder. This is why in our interactions with the administration we encourage them to keep making a strong case for diversity and inclusion and how central it is to the university's mission.

Moving forward at the system level, President DiPietro announced the creation of a part-time position, systemwide Special Advisor to the President on Diversity and Inclusion, now held by Noma Anderson, who also continues as faculty in the UT Health Science Center College for Health Professions in Memphis where she previously served as dean.[172] Ms. Anderson's salary is shared by the University of Tennessee System and the UT Health Science Center. This position, however, is advisory and does not hold the title of Vice President for Diversity and Inclusion, a position previously held by Theotis Robinson.

After assuming her position in July 2016, Noma Anderson, an African American female, conducted a listening tour of the six UT campuses and met with faculty, staff, administrators, and students. She asked the campuses to share both their diversity and inclusion challenges and successes. The results of the tour led to two areas of particular focus: 1) diversity education and 2) communication across the campuses. Due to the decentralized nature of the campuses and the fact that President DiPietro holds each Chancellor responsible for diversity plans, Anderson indicates that campuses want to establish stronger lines of communication regarding their diversity and inclusion efforts. To that end, she is planning a diversity summit as the first major diversity educational objective that will bring the UT campuses together to discuss best practices and diversity-related issues. As a second major objective, she and her team will be working on a diversity strategic plan for the system and will obtain input and feedback from the campuses in the developmental process. As she explains:

> The next step will be to develop a strategic plan with objectives, but our primary objective is to ensure that diversity and inclusion are woven into the system's mission. Our mission is to educate, discover, and connect research, education, service and outreach. We want to make sure that attention is paid to diversity and inclusion in the mission of the system.

Anderson's vision for the future offers promise for continued progress in diversity organizational learning and parallels the observations of former UTK Faculty Senate President, Bruce MacLennan, in terms of the needed alignment of diversity with institutional mission. In her words, "I want to be sure that diversity and inclusion is a part of the mission of what we do. That it's not an afterthought, but it is a foundational piece of what we do across our mission."

Notes

1. Basu, S. (2005). Letting the hydra roam: Attending to diverse forms of diversity in liberal arts education. In M. L. Ouellett (Ed.), *Teaching inclusively: Resources for course, department & institutional change in higher education* (pp. 21–33). Stillwater, OK: New Forums Press.
2. Kezar, A. (2014). *How colleges change: Understanding, leading, and enacting change*. New York: Routledge.
3. Ibid., p. 22.
4. Williams, D. A. (2013). *Strategic diversity leadership: Activating change and transformation in higher education*. Sterling, VA: Stylus.
5. Ibid.
6. Owen, D. (2009). Privileged social identities and diversity leadership in higher education. *The Review of Higher Education*, 32(2), 185–207.
7. Prasad, A. (2001). Understanding workplace empowerment as inclusion. *The Journal of Applied Behavioral Science*, 37(1), 51–69.

8. *Report on faculty diversity and inclusivity in FAS.* (2016). Retrieved August 30, 2016, from http://fassenate.yale.edu/sites/default/files/files/Reports/FAS%20Senate%20-%202016-05-19%20-%20Diversity%20and%20InclusivityFINAL%20copy%202.pdf
9. Chun, E., and Evans, A. (2009). *Bridging the diversity divide: Globalization and reciprocal empowerment in higher education* (ASHE Higher Education Report, Vol. 35, No. 1). San Francisco: Jossey-Bass; Prilleltensky, I., and Gonick, L. S. (1994). The discourse of oppression in the social sciences: Past, present, and future. In E. J. Trickett, R. J. Watts, and D. Birman (Eds.), *Human diversity: Perspectives on people in context* (pp. 145–177). San Francisco: Jossey-Bass.
10. Evans, A., and Chun, E. (2007). *Are the walls really down? Behavioral and organizational barriers to faculty and staff diversity* (ASHE-ERIC Higher Education Report, Vol. 33, No. 1). San Francisco: Jossey-Bass.
11. Roberson, Q. M. (2006). Disentangling the meanings of diversity and inclusion in organizations. *Group & Organization Management*, 31(2), 212–236.
12. Chang, J. (2016). *We gon' be alright: Notes on race and resegregation*, p. 42. New York: Picador.
13. *Making excellence inclusive.* (n.d.). Association of American Colleges & Universities. Retrieved January 25, 2017, from www.aacu.org/making-excellence-inclusive
14. Ibid.
15. *Pathways to diversity and inclusion: An action plan for Brown University.* (2016). Retrieved August 28, 2016, from www.brown.edu/about/administration/institutional-diversity/pathways
16. Faust, D. (2016, September 21). World university rankings 2016–2017: Higher education's diverse mission. *IELTS.* Retrieved January 31, 2017, from www.timeshighereducation.com/world-university-rankings/world-university-rankings-2016-2017-higher-educations-diverse-mission
17. Orfield, G. (2001). Introduction. In G. Orfield (with M. Kurlaender) (Eds.), *Diversity challenged: Evidence on the impact of affirmative action* (pp. 1–29). Cambridge, MA: Harvard Education Publishing Group.
18. For a review of the six primary streams of research on the educational benefits of diversity, see Chun, E., and Evans, A. (2015). *Affirmative action at a crossroads: Fisher and forward* (ASHE Higher Education Report, Vol. 41, No. 4). San Francisco: Jossey-Bass.
19. Bowman, N. A. (2011). Promoting participation in a diverse democracy: A meta-analysis of college diversity experiences and civic engagement. *Review of Educational Research*, 81(1), 29–68.
20. Smith, D. G. (2014). Introduction: The national, international, and institutional context for diversity. In D. G. Smith (Ed.), *Diversity and inclusion in higher education: Emerging perspectives on institutional transformation* (pp. 3–9). New York: Routledge.
21. Ibid.
22. Kennedy, R. (2013). *For discrimination: Race, affirmative action, and the law.* New York: Vintage Books; Moses, M. S. (2006). Toward a deeper understanding of the diversity rationale. *Educational Researcher*, 35(1), 6–11.
23. Myers, S. L., Jr. (1997). Why diversity is a smoke screen for affirmative action. *Change*, 29(4), 24–32; Wise, T. J. (2005). *Affirmative action: Racial preference in black and white.* New York: Routledge.
24. Feagin, J. R. (2006). *Systemic racism: A theory of oppression.* New York: Routledge.
25. Ahonen, P., Tienari, J., Merilainen, S., and Pullen, A. (2014). p. 272. Hidden contexts and invisible power relations: A Foucauldian reading of diversity research. *Human Relations*, 67(3), 263–286. See also Eagleton-Pierce, M. (2016). *Neoliberalism: The key concepts.* New York: Routledge.
26. Feagin. (2006). *Systemic racism.*

27. Chang. (2016). *We gon' be alright*, p. 16.
28. Ibid., p. 31.
29. Owen. (2009). Privileged social identities and diversity leadership in higher education.
30. See for example Collins, P. H. (1993). Learning from the outsider within: The sociological significance of black feminist thought. In J. S. Glazer-Raymo, E. M. Bensimon, and B. K. Townsend (Eds.), *Women in higher education: A feminist perspective* (pp. 45–64). Needham Heights, MA: Ginn Press.
31. Museus, S. D., and Griffin, K. A. (2011). Mapping the margins in higher education: On the promise of intersectionality frameworks in research and discourse. In K. A. Griffin and S. D. Museus (Eds.), *Using mixed methods to study intersectionality in higher education*. Hoboken, NJ: John Wiley & Sons, Inc.
32. Kezar. (2014). *How colleges change.*
33. Levin, J. S. (2000). The practitioner's dilemma: Understanding and managing change in the academic institution. In A. M. Hoffman and R. W. Summers (Eds.), *Managing Colleges and Universities: Issues for leadership* (pp. 29–42). Westport, CT: Bergin & Garvey.
34. Ruben, B. D., De Lisi, R., and Gigliotti, R. A. (2017). *A guide for leaders in higher education: Core concepts, competencies, and tools*. Sterling, VA: Stylus.
35. Birnbaum, R. (1988). *How colleges work: The cybernetics of academic organization and leadership*. San Francisco: Jossey-Bass.
36. Chun, E., and Evans, A. (2012). *Diverse administrators in peril: The new indentured class in higher education*. Boulder, CO: Paradigm Publishers. See also Kezar. (2014). *How colleges change.*
37. Kezar, A. (2001). *Understanding and facilitating organizational change in higher education in the 21st century* (ASHE-ERIC Higher Education Reports, Vol. 28, No. 4). San Francisco: Jossey-Bass. See also Yakoboski, P. J., and Foster, J. E. (2014). Strategic utilization of adjunct and other contingent faculty. *TIAA*. Retrieved August 19, 2016, from www.tiaainstitute.org/public/institute/research/strategic-utilization-of-adjunct-and-other-contingent-faculty
38. Berube, M., and Ruth, J. (2015). *The humanities, higher education, and academic freedom: Three necessary arguments*. New York: Palgrave Macmillan.
39. White, J., and Weathersby, R. (2005). Can universities become true learning organizations? *The Learning Organization* 12(3), 292–298.
40. Hammond, T. H. (2004). Herding cats in university hierarchies: Formal structure and policy choice in American research universities. In R. G. Ehrenberg (Ed.), *Governing Academia* (pp. 91–138). Ithaca, NY: Cornell University.
41. Bateman, O. (2016, October 5). Why the 'safe-space' debate is a problem for adjuncts. *The Atlantic*. Retrieved February 1, 2017, from www.theatlantic.com/education/archive/2016/10/a-power-struggle-inside-safe-spaces/502859/
42. Ibid.
43. Evans and Chun. (2007). *Are the walls really down?*
44. American Association of University Professors. (n.d.). *1966 statement on government of colleges and universities*. Retrieved November 16, 2013, from www.aaup.org/report/1966-statement-government-colleges-and-universities
45. Kezar. (2014). *How colleges change.*
46. Green, D. O., and Trent, W. T. (2005). The public good and a racially diverse democracy. In A. J. Kezar, A. C. Chambers, and J. C. Burkhardt (Eds.), *Higher education for the public good: Emerging voices from a national movement* (pp. 102–124). San Francisco: Jossey-Bass.
47. Slaughter, S., and Rhoades, G. (2004). *Academic capitalism in the new economy: Challenges and choices*. Baltimore: The Johns Hopkins University Press.
48. Kezar. (2014). *How colleges change.*

49. Chun and Evans. (2012). *Diverse administrators in peril.*
50. Williams, D. A. (2014). Organizational learning as a framework for overcoming glass ceiling effects in higher education. In J. F. L. Jackson, E. M. O'Callaghan, and R. A. Leon (Eds.), *Measuring glass ceiling effects in higher education: Opportunities and challenges* (pp. 75–84). San Francisco: Wiley Periodicals.
51. Williams. (2013). *Strategic diversity leadership.*
52. Kezar, A. (2000). Pluralistic leadership: Incorporating diverse voices. *Journal of Higher Education*, 71(6), 722–743.
53. Ibid.
54. Chen, D. (2014). *Achieving diversity in higher education: Faculty leaders' perceptions of culturally responsive leadership.* Unpublished doctoral dissertation, University of Wyoming; Santamaria, L. J., and Santamaria, A. P. (2016). Introduction: The urgent call for culturally responsive leadership in higher education. In L. Santamaria and A. Santamaria (Eds.), *Culturally responsive leadership in higher education: Promoting access, equity, and improvement* (pp. 1–14). New York: Routledge.
55. Chen. (2014). *Achieving diversity in higher education.*
56. Kezar. (2000). Pluralistic leadership.
    Kezar, A., and Carducci, R. (2009). Revolutionizing leadership development: Lessons from research and theory. In A. Kezar (Ed.), *Rethinking leadership in a complex, multicultural, and global environment: New concepts and models for higher education* (pp. 1–38). Sterling, VA: Stylus.
57. Ibid.
58. Kezar. (2000). Pluralistic leadership. See also Kezar and Carducci. (2009). *Revolutionizing leadership development.*
59. Kezar and Carducci. (2009). *Revolutionizing leadership development.*
60. Kezar. (2000). Pluralistic leadership.
61. Cook, B., and Kim, Y. (2013). *The American College President 2012.* Washington, DC: American Council on Education.
62. Bass, B.M., and Avolio, B.J. (1993). Transformational leadership: A response to critiques. In M.M. Chemers and R. Ayman (Eds.), *Leadership theory and research: Perspectives and directions* (pp. 49–80). San Diego, CA: Academic Press. Kezar, A., and Eckel, P. (2008). Advancing diversity agendas on campus: Examining transactional and transformational presidential leadership styles *International Journal of Leadership in Education,* 11(4), 379–405.
63. Berson, Y., Shamir, B., Avolio, B. J., and Popper, M. (2001). The relationship between vision strength, leadership style, and context. *The Leadership Quarterly*, 12(1), 53–73.
64. Kezar and Eckel. (2008). Advancing diversity agendas on campus.
65. Ibid.
66. Ibid.
67. Ibid.
68. Berdahl, J. L., and Min, J. A. (2012). Prescriptive stereotypes and workplace consequences for East Asians in North America. *Cultural Diversity and Ethnic Minority Psychology*, 18(2), 141–152.
69. Chun, E. (2017). The balancing act between governing boards and college or university administration on diversity and inclusion. In R. Thompson-Miller and K. Ducey (Eds.), *Systemic racism: Making liberty, justice, and democracy real*, 79–110. Palgrave MacMillan.
70. Ibid.
71. Schmidt, B. C. (2014). *Governance for a new era: A blueprint for higher education trustees.* Retrieved February 14, 2016, from www.goacta.org/images/download /governance_for_a_new_era.pdf

72. Brimhall-Vargas, M. (2012). The myth of institutionalizing diversity: Structures and the covert decisions they make. In C. Clark, K. Fasching-Varner, and M. Brimhall-Vargas (Eds.), *Occupying the academy: Just how important is diversity in higher education?* (pp. 85–95). Lanham, MD: Rowman and Littlefield.
73. Ibid.
74. Jackson, B. W. (2012). Stories from the Chief Diversity Officer frontlines. In C. Clark, K. Fasching-Varner, and M. Brimhall-Vargas (Eds.), *Occupying the academy: Just how important is diversity work in higher education?* (pp. 21–22). Lanham, MD: Rowman & Littlefield.
75. Nixon, M. L. (2013). *Women of color chief diversity officers: Their positionality and agency in higher education institutions.* Retrieved January 17, 2017, from https://digital.lib.washington.edu/researchworks/bitstream/handle/1773/23632/Nixon_washington_0250E_11553.pdf?sequence=1
76. Nixon, M. L. (2016). Experiences of women of color university chief diversity officers. *Journal of Diversity in Higher Education.* Retrieved October 13, 2017 from http://psycnet.apa.org/doiLanding?doi=10.1037%2Fdhe0000043
77. Anderson, L. (2012). Deconstructing hope: A chief diversity officer's dilemma in the Obama era. In C. Clark, K. Fasching-Varner, and M. Brimhall-Vargas (Eds.), *Occupying the academy: Just how important is diversity work in higher education?* (pp. 50–60). Lanham, MD: Rowman & Littlefield.
78. King, J., and Gomez, G. G. (2008). *On the pathway to the presidency: Characteristics of higher education's senior leadership.* Washington, DC: American Council on Education.
79. Ibid.
80. Nixon. (2013). *Women of color chief diversity officers.*
81. Owen, D. S. (2009). Privileged social identities and diversity leadership in higher education.
82. Nixon. (2013). *Women of color chief diversity officers.*
83. Ibid., p. 113.
84. Ibid.
85. Williams, D. A., and Wade-Golden, K. C. (2013). *The chief diversity officer: Strategy structure, and change management.* Sterling, VA: Stylus, 32.
86. McMurtrie, B. (2016, April 7). 2 new diversity deans take on ivy league challenges. *The Chronicle of Higher Education.* Retrieved August 9, 2016, from http://chronicle.com/article/2-New-Diversity-Deans-Take-On/236013
87. *Stanford: Graduate Diversity.* Retrieved December 25, 2016, from https://graddiversity.stanford.edu/about/contacts
88. Clark, C., Fasching-Varner, K. J., and Brimhall-Vargas, M. (2012). Occupying academia, reaffirming diversity. In C. Clark and K. Fasching-Varner (Eds.), *Occupying the academy: Just how important is diversity work in higher education?* (pp. 1–20). Lanham, MD: Rowman & Littlefield Publishers.
89. Ibid.
90. Kezar, A., and Lester, J. (2009). Supporting faculty grassroots leadership. *Research in Higher Education*, 50(7), 715–740.
91. Hart, J. (2009). Creating faculty activism and grassroots leadership: An open letter to aspiring activists. In A. Kezar (Ed.), *Rethinking leadership in a complex, multicultural, and global environment: New concepts and models for higher education* (pp. 169–184). Sterling, VA: Stylus.
92. Kezar and Lester. (2009). Supporting faculty grassroots leadership.
93. Kezar. (2007). Tools for a time and place.

94. Smith, D. G. (2014). Introduction: The national, international, and institutional context for diversity. In D. G. Smith (Ed.), *Diversity and inclusion in higher education: Emerging perspectives on institutional transformation* (pp. 3–9). New York: Routledge.
95. Kezar. (2014). *How colleges change*.
96. Wilson, J. L., Meyer, K. A., and McNeal, L. (2012). Mission and diversity statements: What they do and do not say. *Innovative Higher Education*, 37(2), 125–139.
97. Edmondson, A. C. (2002). The local and variegated nature of learning in organizations: A group-level perspective. *Organization Science*, 13(2), 128–146; Kezar, A. (2005). What campuses need to know about organizational learning and the learning organization. In A. J. Kezar (Ed.), *Organizational learning in higher education* (pp. 7–22). San Francisco: Jossey-Bass.
98. Simon, H. A. (1991). Bounded rationality and organizational learning. *Organization Science*, 2(1), 125–134.
99. Chun, E., and Evans, A. (2009). *Bridging the diversity divide: Globalization and reciprocal empowerment in higher education* (ASHE-ERIC Higher Education Reports, Vol. 35, No. 1). San Francisco: Jossey-Bass. See also Williams. (2013). *Strategic diversity leadership*.
100. Argyris, C. (1997). Learning and teaching: A theory of action perspective. *Journal of Management Education*, 21(1), 19–26.
101. Ibid.
102. Brown, A. D., and Starkey, K. (2000). Organizational identity and learning: A psychodynamic perspective. *The Academy of Management Review*, 25(1), 102–120, p. 5.
103. Argyris, C. (1990). *Overcoming organizational defenses: Facilitating organizational learning*. Boston: Allyn & Bacon.
104. Argyris, C. (1993). Education for leading-learning. *Organizational Dynamics*, 21(3), 5–17. Argyris. (1997). Learning and teaching; Argyris, C., and Schon, D. A. (1996). *Organizational learning II: Theory, method, and practice*. Reading, MA: Addison-Wesley.
105. Williams. (2013). *Strategic diversity leadership*.
106. Argyris. (1993). Education for leading-learning. See also Argyris. (1997). Learning and teaching. Argyris and Schon. (1996). *Organizational learning II*.
107. Ibid.
108. Yeung, A. K., Ulrich, D., Nason, S. W., and Von Glinow, M. A. (1999). *Organizational learning capability: Generating and generalizing ideas with impact*. New York: Oxford University Press, 11.
109. Ibid., 13.
110. Ibid.
111. Edmondson. (2002). The local and variegated nature of learning in organizations. Kezar, A. (2005). What campuses need to know about organizational learning and the learning organization. In A. J. Kezar (Ed.), *Organizational learning in higher education* (pp. 7–22). San Francisco: Jossey-Bass.
112. Yeung, Ulrich, Nason, and Von Glinow. (1999). *Organizational learning capability*.
113. Ibid. See also Kezar. (2005). What campuses need to know about organizational learning and the learning organization.
114. Brown and Starkey. (2000). Organizational identity and learning. See also Smith, D. G., and Parker, S. (2005). Organizational learning: A tool for diversity and institutional effectiveness. In A. J. Kezar (Ed.), *Organizational learning in higher education* (pp. 113–125). San Francisco: Jossey-Bass.
115. Smith and Parker. (2005). Organizational learning.
116. Argyris, C., Bellman, G. M., Blanchard, K., and Block, P. (1994). The future of workplace learning and performance. *Training & Development*, 48(5), 36–47.

117. Evans, A., and Chun, E. (2012). *Creating a tipping point: Strategic human resources in higher education* (ASHE Higher Education Report, Vol. 38, No. 1). San Francisco: Jossey-Bass.
118. Edmondson. (2002). The local and variegated nature of learning in organizations.
119. Argyris. (1997). Learning and teaching.
120. Edmondson, A. C. (2004). Psychological safety, trust, and learning in organizations: A group-level lens. In R. M. Kramer and K. S. Cook (Eds.), *Trust and Distrust in organizations: Dilemmas and approaches* (pp. 239–272). New York: Russell Sage Foundation.
121. Feagin, J. R. (2016, October 3). Personal communication.
122. Evans and Chun. (2012). *Creating a tipping point.*
123. Burnes, B. (2009). Kurt Lewin and the planned approach to change: A reappraisal. In W. W. Burke, D. G. Lake, and J. W. Paine (Eds.), *Organization change: A comprehensive reader* (pp. 226–254). San Francisco: John Wiley & Sons.
124. Ibid.
125. Edmondson. (2004). Psychological safety, trust, and learning in organizations.
126. White and Weathersby. (2005). Can universities become true learning organizations?
127. Smith and Parker. (2005). Organizational learning.
128. *QuickFacts: Shelby County, Tennessee.* (n.d.). Retrieved August 25, 2016, from www.census.gov/quickfacts/#table/PST045215/47157
129. UT desegregation timeline. (n.d.). Retrieved August 25, 2016, from http://achieve.utk.edu/timeline.shtml
130. *Theotis Robinson Jr.* (n.d.). The University of Tennessee Knoxville. Retrieved August 25, 2016, from http://trailblazer.utk.edu/2014-2015/theotis-robinson/
131. MacTaggart, T. J. (2004). The ambiguous future of public higher education systems. In W. G. Tierney (Ed.), *Competing conceptions of academic governance: Negotiating the perfect storm* (pp. 104–136). Baltimore, MD: Johns Hopkins University Press.
132. *Tennessee higher education commission: Policy manual.* (n.d.). Retrieved September 2, 2016, from www.tn.gov/assets/entities/thec/attachments/Policy_Manual_as_revised_01-29-15.pdf
133. *Welcome to the UTK 2015*–16 *fact book.* (n.d.). The University of Tennessee Knoxville. Retrieved August 31, 2016, from https://oira.utk.edu/factbook
134. *Continuing our journey to the top: Past progress and future opportunities a five-year assessment of Vol Vision.* (2015). Retrieved September 1, 2016, from http://top25.utk.edu/wp-content/uploads/sites/11/2015/09/VolVision2020ReportDraft.pdf
135. *Vol vision: The pursuit of top 25.* (2011). Retrieved September 1, 2016, from http://trace.tennessee.edu/cgi/viewcontent.cgi?article=1003&context=utk_chanstrategic
136. *Continuing our journey to the top: Past progress and future opportunities a five-year assessment of Vol Vision.* The University of Tennessee, Knoxville. (2015). Retrieved October 13, 2017 from http://top25.utk.edu/wp-content/uploads/sites/11/2015/09/VolVision2020ReportDraft.pdf. See also *Vol Vision 2020: Journey to the top.* (2016). The University of Tennessee, Knoxville. Retrieved October 13, 2017 from http://top25.utk.edu/wp-content/uploads/sites/11/2016/04/Vol-Vision-2020.pdf
137. Kingkade, T. (2014, March 12). Tennessee lawmakers aim to punish university for 'sex week' by cutting back public debate. *The Huffington Post.* Retrieved September 1, 2016, from www.huffingtonpost.com/2014/03/12/tennessee-lawmakers-sex-week_n_4944060.html
138. Ibid.
139. Jaschik, S. (2016, April 22). Defunding diversity. *Inside Higher Ed.* Retrieved August 31, 2016, from www.insidehighered.com/news/2016/04/22/both-houses-tennessee-legislature-vote-bar-use-state-funds-university-diversit. Kumar, A. (2015, September 12).

University of Tennessee president withdraws transgender pronouns 'ze,' 'zir' advisory, saying controversy 'like nothing I've seen.' *Christian Post*. Retrieved September 1, 2016, from www.christianpost.com/news/university-of-tennessee-president-withdraws-transgender-pronouns-ze-zir-advisory-saying-controversy-like-nothing-ive-seen-145154/

140. Pride Center. (n.d.). *The University of Tennessee Knoxville*. Retrieved September 1, 2016, from http://pridecenter.utk.edu/
141. Ibid.
142. Jaschik, S. (2015, September 8). Fear of new pronouns. *Inside Higher Ed*. Retrieved August 31, 2016, from www.insidehighered.com/news/2015/09/08/u-tennessee-withdraws-guide-pronouns-preferred-some-transgender-people
143. Ibid.
144. Starnes, T. (2015, December 3). University: Your holiday party cannot be a Christmas party in disguise. *Fox News*. Retrieved August 31, 2016, from www.foxnews.com/opinion/2015/12/03/university-your-holiday-party-cannot-be-christmas-party-in-disguise.html
145. Ibid.
146. Jaschik, S. (2015, December 7). War on Christmas? On inclusivity? *Inside Higher Ed*. Retrieved August 31, 2016, from www.insidehighered.com/news/2015/12/07/furor-over-holiday-party-advice-u-tennessee-grows-there-and-spreads
147. UT Knoxville reiterates commitment to diversity and inclusion, takes steps to move forward. (2015). *The University of Tennessee Knoxville*. Retrieved August 31, 2016, from http://tntoday.utk.edu/2015/12/08/ut-knoxville-reiterates-commitment-diversity-inclusion-takes-steps-move/
148. Ibid.
149. Boehnke, M. (2015, December 7). Lawmaker says he'll draft bill to defund UT diversity office. *Knoxville News Sentinel*. Retrieved August 31, 2016, from www.knoxnews.com/news/local/lawmaker-says-hell-draft-bill-to-defund-ut-diversity-office-ep-1403967097–361255691.html
150. Schallhorn, K. (2015). *State lawmakers consider pulling university funding after discovering how much money is being spent on 'diversity.'* Retrieved August 31, 2016, from www.theblaze.com/stories/2015/12/08/state-lawmakers-consider-pulling-university-funding-after-discovering-how-much-money-is-being-spend-on-diversity/
151. *Tennessee this week—December 13, 2015*. (2015). Retrieved August 30, 2016, from http://wate.com/2015/12/13/tennessee-this-week-december-13-2015/
152. Ibid.
153. Kingkade, T. (2016, January 12). Republicans in Tennessee want state investigation of university diversity offices: The diversity office controls less than 0.25 percent of the University of Tennessee's annual budget. *The Huffington Post*. Retrieved August 31, 2016, from www.huffingtonpost.com/entry/university-of-tennessee-diversity_us_56942dc7e4b086bc1cd4ee5a
154. Ibid.
155. Loewen, R. A. (2015, September 9). *Funding for UT diversity programs*. Retrieved August 31, 2016, from http://senate.utk.edu/wp-content/uploads/sites/16/2015/12/Response-to-Legislative-Budget-Office5.pdf
156. Locker, R. (2016, May 20). University of Tennessee diversity funding bill allowed to become law. *The Tennessean*. Retrieved August 31, 2016, from www.tennessean.com/story/news/politics/2016/05/20/university-tennessee-diversity-funding-bill-allowed-become-law/84650208/
157. Ibid.

158. Jaschik. (2016). Defunding diversity.
159. Duda, C. (2016). *Hundreds march at UT Knoxville to protest diversity funding cuts.* Retrieved August 31, 2016, from www.knoxmercury.com/2016/04/19/hundreds-march-ut-knoxville-protest-diversity-funding-cuts/
160. *UT diversity matters.* (n.d.). Retrieved August 31, 2016, from https://assets.documentcloud.org/documents/2701237/Diversity.pdf
161. Ibid.
162. North, J. (2016). *UT's diversity office dismantled, chief takes job at University of Washington.* Retrieved August 31, 2016, from www.wbir.com/news/local/education/uts-diversity-head-takes-job-at-university-of-washington/207323628
163. *Tennessee lawmaker plans to defund UT's office of diversity.* (2015). Retrieved August 31, 2016, from http://wate.com/2015/12/07/university-of-tennessee-faculty-support-cheek-hall-on-workplace-holiday-party-guidelines/
164. Pignolet, J. (2016, April 21). DiPietro talks of diversity funding and guns on campus. *The Commercial Appeal.* Retrieved August 31, 2016, from www.commercialappeal.com/news/schools/dipietro-talks-of-diversity-funding-and-guns-on-campus—3101f512–79d3–394e-e053–0100007ff14c-376622311.html
165. Ibid.
166. Ibid.
167. Rau, N., and Wadhwani, A. (2016, July 6). Tennessee settles sexual assault suit for $2.48 million. *The Tennessean.* Retrieved August 31, 2016, from www.tennessean.com/story/news/crime/2016/07/05/tennessee-settles-sexual-assault-suit-248-million/86708442/. Slaby, M. J. (2016, June 21). Jimmy Cheek's return to classroom comes after rocky year at University of Tennessee. *Knoxville News Sentinel.* Retrieved August 31, 2016, from www.knoxnews.com/news/local/cheek-retirement-comes-after-rocky-year-at-ut-35c87f1e-4aad-65ae-e053–0100007f0f18–383882531.html
168. *Diversity at Tennessee: Course syllabi and diversity.* (2016). Retrieved August 31, 2016, from http://senate.utk.edu/diversity-at-tennessee/
169. Duda, C. (2016, June 22). *Faculty and students seek path forward for UT pride center.* Retrieved August 29, 2016, from www.knoxmercury.com/2016/06/22/faculty-students-seek-path-forward-ut-pride-center/
170. Ibid.
171. *Strategies and tactics for recruiting to improve diversity and excellence at the University of Tennessee.* (n.d.). Retrieved August 31, 2016, from http://stride.utk.edu/wp-content/uploads/sites/14/2014/12/STRIDE-Presentation-for-Search-Committees-2015.pdf
172. Duda, C. (2016, June 24). *In wake of cuts, UT appoints part-time, systemwide diversity advisor.* Retrieved August 22, 2016, from www.knoxmercury.com/2016/06/24/wake-cuts-ut-appoints-part-time-system-wide-diversity-advisor/. See also *Dr. Norma Anderson.* Office of the President, University of Tennessee. Retrieved March 20, 2017, from http://president.tennessee.edu/biography/noma-anderson

3

A FRAMEWORK FOR SYSTEMATIC DIVERSITY ORGANIZATIONAL LEARNING

> Our general thesis is that universities in the United States have often been somewhat poorly managed and that improved leadership and better management of the university is not just desirable but now essential.
>
> Sohvi Leih and David Teece, 2016, p. 2[1]

> Management is in desperate need of an overhaul. . . . Management is the technology of human accomplishment. We need to reinvent it, root and branch.
>
> Gary Hamel[2]

The reinvention of campus diversity leadership practices clearly remains an evolutionary, unfinished, and ongoing process. As a result, in this chapter we introduce the Inclusive Excellence (IE) Change model as an organizing framework for diversity that re-centers institutional diversity strategy on student learning outcomes. Despite the need to prepare students for careers and citizenship in a global workforce and society, scholars find little evidence that colleges and universities have coordinated a systemic effort to connect diversity experiences with courses for student learning or to implement the necessary practices and policies "to make diversity work."[3] When administrators have led diversity efforts without substantive input from faculty, staff, and students, these efforts have failed to crystallize an overarching framework that ties institutional policies and practices to the attainment of diversity learning outcomes.[4] Only in recent years has scholarly attention been given to the politics surrounding diversity and the challenges faced by leaders in moving the diversity agenda forward.[5]

The emphasis of the IE model on the centrality of the student experience in creating a welcoming and inclusive campus environment requires consideration of how to effect cultural change and the obstacles to such change. Following a discussion of leading theories of cultural change, we delineate the elements of campus climate and culture that can neutralize or derail diversity. We conclude the chapter by exploring the primary phases of a diversity culture shift and the hallmarks of each phase.

The Inclusive Excellence Change Model

Due to the frequent pushback over whether diversity trumps quality, the concept of Inclusive Excellence was introduced by the American Association of Colleges and Universities (AAC&U) as an organizing paradigm for the diversity and inclusion. The framework of Inclusive Excellence (IE) emphasizes the ways in which diversity and quality bond together to form a stronger and more durable alloy through four key elements: 1) enhancing the social and intellectual development of students, 2) recognizing and attending to the cultural differences that students bring to the educational environment, 3) purposeful utilization of institutional resources in support of student learning, and 4) building a welcoming campus community.[6] All four of the components of Inclusive Excellence focus on student development and student engagement. As such the IE model addresses the need for institutional diversity policies and goals to be redirected toward students within the context of student learning outcomes and an inclusive campus culture.

From an institutional perspective, IE has been deployed as an overall change model that infuses policies and practices and provides a touchstone for inclusion efforts.[7] A number of colleges and universities across the nation have introduced this model for measuring progress in diversity, equity, excellence, and inclusion. The process of operationalizing this student-centered framework throughout the multiple dimensions of a campus ecosystem presents major challenges.

Yet Yekim, the Afro-Latino CDO cited earlier, describes the significant limitations of the Inclusive Excellence framework in terms of its non-specificity, lack of grounding in the realities of power and privilege, and the potential for use as a rhetorical reference point without a concrete material commitment:

> Even the idea of Inclusive Excellence is not really defined in a way that gives concrete understanding to the work that has to be done. . . . when I think about the contribution of my office to Inclusive Excellence, I think that at times we still get caught up on the end result, right? To assure that every student is successful and graduates and leaves with a critical understanding of how they are going to be contributing citizens to a just world. But I think that even that doesn't tackle power; it doesn't tackle privilege from a place of structure and content. And even Inclusive Excellence falls short, because very few people are willing to signify and demarcate the 'dominant discomfort' that needs to happen in order to address the issues that lead to more authentic inclusive excellence.
>
> When I think about all the institutions . . ., first, most of the institutions define Inclusive Excellence with the general framework but still in their own way. But second, very few have a real material commitment to it. Without the material commitment, you cannot rely on what is already in place and tagging onto those roles, instead of actually creating a commitment that infiltrates the structure. Right now, I believe, even with all the work done, I believe the work is done in a way that doesn't address power. Even if we have critical conversations across higher ed campuses and academia, nobody is addressing the resource part. As long as the resource part is not addressed, then Inclusive Excellence is going to be limited to the preferencing of ideas, not necessarily the centralizing of the work that needs to happen structurally.

This critique calls for amplification of IE's tenets to include the empowerment of nondominant groups through practices that give voice and participation to marginalized individuals. Although a number of institutions have adopted Inclusive Excellence as the change model for diversity progress, the model can be implemented as a rhetorical reference point without the concomitant recognition of the persistence of structures, practices, and processes that privilege the dominant group in power and the need for concrete material commitment to the change process.

For example, the power imbalance within institutions of higher education is reflected in the cultural masculinity resulting from the predominance of white males in top leadership roles compared to the often symbolic representation of minority chief diversity officers.[8] From this vantage point, the reproduction of asymmetrical power structures and hierarchically based systems of inequality within higher education has preserved and reinforced the marginalization and exclusion of minoritized groups whether among students, faculty, administrators, or staff.

In this context, Larry Burnley, Vice President for Diversity and Inclusion at the University of Dayton, views the IE change model as disruptive since it necessarily involves an institutional self-critique:

> This notion of Inclusive Excellence is a very disruptive process. While not the goal, IE is very conflict-oriented, because we are really talking about rearranging and deconstructing structurally and systemically embedded systems of power and privilege that only exist to protect and reproduce themselves. Deconstruction and redefining the notion of excellence is very threatening to those who are vested in maintaining the status quo. The hope, of course, is the tension and conflict resulting from work aimed at the individual and structural embodiment of IE is constructive, moving us toward sustainable equitable outcomes and not destructive. IE is intended to make a significant contribution toward the common good!

Burnley, an African American male, identifies the need to deconstruct and reconceptualize even the way excellence itself has been viewed:

> IE begins with the assumption that in order to be excellent, one must be inclusive. In other words, everyone counts. . . . So to do that, institutions are called to begin to ask the question: what is it about our practice and policy that results in marginalization and exclusion of voices in our community based on identity, based on

> how they are located, economically, socially, politically? And it also begins with the need to rethink, reconceptualize, or deconstruct dominant conceptualizations of excellence itself. . . . How do we define that? How is that informed by where we are positioned in terms of power, privilege, identity, socioeconomic location? What are the assumptions that are being made in how we understand excellence? The whole notion of excellence has to be reconsidered.

Given these considerations, the IE change model cannot be understood as a checklist of certain percentages of minoritized students or faculty or the number of programs relating to diversity.[9] Rather it must be understood as a process leading toward *enhanced learning* rather than an outcome.[10] IE opens the door to *enhanced organizational learning*. In practice, then, the IE model will require reframing and deframing of white racial, masculine, heterosexist, ableist frameworks historically reinforced by practices that perpetuate privilege through informal practices, processes, and networks. In fact, the normalization of contemporary racism, sexism, and heterosexism through covert and subtle forms of discrimination have been shown to affect the day-to-day experiences of minoritized students, faculty, and administrators in higher education, leading to discriminatory actions.[11] As Burnley suggests, the organizational learning process requires reframing of excellence and understanding the ways in which institutions have perpetuated and reproduced social inequality.

Consider the perspective of Lisa Coleman, Chief Diversity Officer and Special Assistant to the President at Harvard University, an African American female, who views Inclusive Excellence as fundamentally about learning and creating a culture of learning about diversity and inclusion:

> I also think Inclusive Excellence ultimately is about *learning* and creating a *culture of learning* versus a culture of experts. In the academic world we have a lot of experts, meaning faculty and core administrators who also need to understand the diversity and inclusion and belonging space.

In September 2017, Coleman assumed the position of senior vice president for global inclusion, diversity, and strategic innovation and inaugural chief diversity officer at New York University. From a similar vantage point, Teresa, an Asian American CDO at an eastern religiously-affiliated university who identifies as queer and bisexual, describes her institution's reframing of diversity from a multicultural model or affirmative action model to a model focused on the learning process in support of Inclusive Excellence:

> This is really about integrating inclusion into the entire learning process. . . . This is something that all members of the campus community can and should be engaged in. . . . That I think is the work of Inclusive Excellence: it's integrating it into the enterprise and then into the structures. And Inclusive Excellence also is about looking at the cultural assumptions, naming those, asking if they work for everyone, if they don't how do we create that next level down. So it goes beyond, "Oh, we are going to hire, we're going to advertise in the Chronicle and with some affinity groups and we're going to hire a person of color." But it really does get to the behavioral elements, how do people interact, how do people feel, the psychological, and then also the systemic, how do we sustain the work?

Dennis Mitchell, Vice Provost for Faculty Diversity and Inclusion at Columbia University, views Inclusive Excellence as a moving target with the ultimate aim of bringing diverse talent to the campus in order to enhance scholarship and areas of inclusion that impact the learning experience for students:

> It's a moving target in my opinion. I think that we are working towards a future where in a perfect world, all of us who work in the field of diversity work towards a time when we no longer have to do this work, right? In this ideal world we could get to, that's where we would like to be. The goal line keeps moving on us and that's a bit of a challenge, but I think we want to keep the scholarly aspect of this alive; we're in higher education, so we want to make sure that we bring the finest scholars and the areas of inclusion that are the most topical to our campus and keep that elevated at all times and continue along that line.

With the emphasis given by these institutional leaders to the integral link of the IE change model to organizational learning, we now examine prominent theories of cultural change that will advance the attainment of a diversity culture shift.

Prominent Theories of Cultural Change

The research literature offers a number of theories for organizational change in higher education that are applicable to the process of diversity transformation. A study of six institutions engaged in change processes over a four-year period identified a relationship between culture and change, and underscored the importance of examining multiple cultural layers and applying culturally appropriate strategies to the change process.[12] Most notably, Adrianna Kezar identifies six frameworks or schools of thoughts which can be melded or used simultaneously to inform the change process.[13] These frameworks are described briefly here:

1. **Scientific Management** or planned change models that draw upon organizational development, strategic planning, reengineering, and professional development to produce first-order changes in structures and processes;
2. **Evolutionary** models that take a cue from biological evolution and emphasize context, external environment and the interaction of interdependent systems and structures in an adaptive, yet largely unplanned process;
3. **Social Cognition** theories that address sensemaking and organizational learning through changes in individual thought processes and mental models;
4. **Cultural** theory which focuses on deep or pervasive change and emphasizes organizational context, mission and history as well as the multiple levels of organizational culture;
5. **Political** models that emphasize differing agendas and interests that require bargaining, negotiation, and influence in the change process; and
6. **Institutional or Neo-institutional** theory that takes into account the unique nature of educational institutions, emphasizes isomorphism or how institutions have become more similar in character

over time and views the broader field of which individual institutions are a part.

Since our focus is on leadership approaches that foster diversity culture change, we primarily draw upon ideas and perspectives from cultural and social cognition theories of change. These theories emphasize the centrality of organizational learning in shifting mindsets, underlying assumptions, and norms in support of diversity. Nonetheless, given the importance of institutional context, themes from political and institutional models can help diversity leaders address planned, systematic change that is more characteristic of scientific management models. Different aspects of these theoretical approaches can be combined in a cohesive, multifaceted strategy for diversity transformation.

With these theoretical approaches in mind, we turn our focus to power of culture in higher education and the ways in which culture can promote, neutralize, or stifle diversity organizational learning.

Hidden Forces of Cultural Resistance

Although an organic leadership strategy combines academic and administrative leadership with the collaborative input of faculty, staff, and students, organizational culture plays a critical role in the process of cultural change. As Edgar Schein indicates, leadership and organizational culture are intertwined and are two sides of the same coin.[14] From this perspective, "the only thing of real importance that leaders do is to create and manage culture."[15] Leaders can destroy culture or understand, work within, and change cultures.[16] In essence, leaders and culture work in a reciprocal relationship in which culture influences leadership and leadership affects culture.[17]

The power of institutional culture cannot be underestimated. Schein observes that culture is the deepest part of an organization and is often less tangible and visible. As such, culture reflects those elements of an organization that are the "most stable and least malleable."[18] Further, culture encapsulates the behavioral, emotional, and cognitive aspects of a group's psychological function.[19] Leaders that overlook the powerful undercurrents of culture can misgauge the campus receptivity to change and mismatch strategies with institutional readiness resulting in frustration and alienation of some campus constituencies, increased turnover, and a delay in the progress of diversity initiatives.[20]

The construct of culture has been described in metaphorical terms as a tapestry or a web. As an amorphous yet influential force, it guides the behaviors of individuals and groups and provides a frame of reference for interpretation of events.[21] Drawing on the perspectives of the anthropologist, Clifford Geertz, who viewed culture as a web spun by humans, Tierney describes the culture of a university as a web of interconnections.[22] The study of institutional culture then not only involves the natural laws and structure of the web, but also how the web is interpreted by institutional actors.[23]

Take the perspective of Antonio, a male Hispanic CDO of an eastern research university, who indicates that the most significant barrier to diversity is adherence to the status quo. He refers to the frequently uttered mantra, "We've always done it this way; it has always worked for us." Given the decentralized structure of universities and colleges and the resulting ambiguities of power and purpose, changing the culture of an institution requires overcoming high inertia.[24] According to social theorist Joe Feagin, social inertia sustains forces of social oppression such as resource inequalities or hierarchical norms unless met with a major, unbalancing force.[25] In light of the pushback resulting from the desire to retain the status quo, Madeleine, an African American director of faculty diversity at a private eastern liberal arts college, emphasizes the slow process of cultural change at her institution:

> it ebbs and flows. I think we're like most predominantly white institutions in that we don't fully get it, we don't engage in a way that's always productive. I feel like it's a piecemeal patchwork of trying to solve and find easy answers to things that are really not solving the long-term issues. I think that at any institution it is more about image more than substance, you can count off what we are doing, but are we being effective? . . . I think there is a lot less lip service to diversity now, although part of that is still there. I think we are more committed to really trying to change some of the culture in some ways, but again there are still various levels of resistance.

Madeleine identifies the subtle, bureaucratic ways in which resistance to diversity culture change takes shape:

> Resistance comes in the form of stalling, in my mind. I was told to write proposals, then rewrite proposals, and tweaked it, do this, get this persons approval. . . . Things took years; little things took years to get anywhere. Major changes just don't go anywhere, they get sucked into the administrative process; we are very focused on process, which is important. . . . But I think that at my institution perhaps at many institutions, we use the excuse of process to never make any change or never do anything. We change the process, and there are more hurdles.

The normative culture of an institution defines what is appropriate, valued, and permissible in terms of behavioral interactions, informal practices, and formal organizational processes. As Teresa, the Asian American CDO cited earlier, explains:

> I don't get the sense that the campus culture is talked about very much here. . . . so I think that's a big part of this: naming the culture, naming who feels comfortable in this culture and then starting to shift that and create some new norms.

Through the interweaving of systems and practices beneath an apparently egalitarian veneer, institutional culture can perpetuate stereotypical ideas and biased attitudes that reinforce the stigmatization of certain groups.[26] Such practices include the naming of realities by the dominant group in power who controls institutional language and the status quo.[27] The hegemonic process of cultural domination overrules the perspectives of the disempowered.[28] Due to the racialized terrain of American society, social norms, roles, language forms, and even preferred modes of sociopolitical thinking are white-generated and white-imposed.[29]

As a result, a diversity culture shift involves bridging the gap between the espoused values of an institution and the predominant norms, assumptions, and traditions that undergird behaviors, actions, and practices. In essence, transforming a system means everything must be reinvented, including new frames, new mindsets, and new infrastructures.[30]

Diversity transformation necessarily invokes the theoretical, psychological, and structural aspects of the change process. Each of these dimensions has the potential for subverting diversity change. For example, little change will likely occur when the CDO position is added to the administrative hierarchy but is disempowered in relation to the traditional organizational structures that reflect the institution's underlying interests.[31]

Sheer demographic diversity on a college campus does not mean that an institution is inclusive in its programs, practices, and the day-to-day experiences of faculty, administrators, staff, and students. The campus climate and culture for diversity is affected by the interplay between external forces such as the laws and policies of governmental agencies and the internal dynamics of an institution.[32] Internal forces shaping the campus climate for diversity include the institution's 1) historical legacy including mission, policies, geographic location, and resistance to desegregation; 2) structural or compositional diversity of faculty, administrators, staff, and students; 3) psychological climate that involves perceptions of discrimination and tensions arising from diversity; and 4) the behavioral dimension that encompasses social interactions across difference, intergroup relations, and classroom diversity.[33] A fifth dimension of the campus climate for diversity identified by Milem, Chang, and Antonio (2005) is organizational/structural diversity that focuses on the structures and decision-making processes that guide day-to-day work on campuses.[34]

Scholars have designated the *"campus racial culture"* to address the ways in which institutional history, mission, norms, values, physical settings, and underlying beliefs shape the experiences of different racial and ethnic groups on campus.[35] The concept of *campus racial climate* is a more specific construct developed to address the experiences of minoritized students on predominantly white campuses and their experiences of isolation and discriminatory treatment.[36] The racial culture can serve to reinforce the views of the dominant group, while oppressing or fail to recognize the perspectives and cultural backgrounds of racial and ethnic minority populations.[37]

To broaden this framework, the concept of a "campus diversity culture" would address the development of a culture of inclusion for members of all nondominant groups. Such groups include not only racial and ethnic minorities, but also LGBT, disabled, veterans, and socioeconomically disadvantaged members of the campus community. Such a concept helps diffuse the concept of "fit" or the perception that certain people fit in an environment while others do not based on deeply embedded social stereotypes. Within the context of higher education, women and minorities are often marginalized due to the notion of being a "weak fit."[38]

The existence of contrasting subcultures and microclimates within the university remains problematic in the effort to create an inclusive campus culture aligned with institutional mission and values. Ben Reese, vice president of the Office for Institutional Equity at Duke University and Duke University Health System, describes the difficulty of generalizing about organizational culture:

> It's hard because the culture of the public policy school is different than the culture of the undergraduate campus and is radically different from the culture of the surgery department. We are so decentralized.

Take, for example, the hostile microclimate in three academic departments at Smith College chronicled in a case study in which female faculty members reported bullying and harassment in several instances by their senior male counterparts.[39] Such microclimates are symptomatic of the isolation of members of nondominant groups within segregated white, masculine networks.[40] In this regard, Catherine, the faculty diversity dean at an eastern liberal arts college cited earlier, observes:

> I think some of our departmental cultures may also be a barrier. In our institution, traditionally academic departments have functioned pretty independently, so each department has its own way of doing things, its own practices and some of those ways of doing things and some of those practices can stand in the way of us becoming a much more diverse, equitable, and inclusive place.

Phases of a Diversity Culture Shift

Several different frameworks have been proposed for understanding the successive stages of diversity culture change. Some researchers suggest that cultural transformation occurs in three progressive phases, a paradigm that can be applied to diversity change.[41] For example, a study of 27 experienced presidents indicates that the process of diversity culture change and organizational learning takes place in the following three stages:[42]

1. *Mobilization* during which diversity vision and direction are set, priorities established, and support systems are created. In this phase, the primary focus is on student access with few conversations about diversity. Diversity programs and a campuswide diversity commitment are absent or present only in nascent stages. Educational processes include attention to behavioral barriers to diversity.
2. *Implementation* in which momentum is built through more systemic support systems as well as rewards and incentives. In this phase, clear rhetoric about diversity exists as well as committed supporters and diversity efforts are less compartmentalized. Although leaders and programs do not work as an integrated whole, they are coordinated loosely. Cultural norms and behaviors reflect receptivity to diversity.
3. *Institutionalization* during which leaders act as cultural agents in building consensus around values and ensuring the durability of the change process. In this phase a diversity agenda is institutionalized, accountability mechanisms are in place, and the focus has shifted from access and retention to outcomes and success. Normative consensus about the value of diversity is reflected in interactions and processes.

In a similar vein, Yolanda Moses describes the first phase of mobilization as characterized by talk relating to diversity including reference in institutional documents without a link to actual practices.[43] In the second phase, institutions have many programs, but these often operate in isolation as "random acts of diversity" without clarity as to how the programs relate to organizational mission.[44] The third phase represents "aspirational diversity" in which the shift occurs leading to institutional transformation.

The recent emergence of diversity mapping methodology provides a comprehensive, research-based framework for gauging progress in diversity and inclusion. This methodology, which has been implemented at more than 20 universities, examines the integration of diversity work on a campus in organizational dimensions that include student learning objectives, curriculum, budget allocations, and policies.[45] Led by researcher Rona Halualani, the framework offers a practice that changes the culture through sustainable efforts.[46] Following is a synopsis of the characteristics of each evolutionary phase of diversity change:[47]

First order—establishes the groundwork for diversity and inclusive excellence as reflected in an institutional commitment to diversity;

Second order—reflects demonstrated commitment to diversity through actions, interventions, and programs

Third order—evidence of sustained, repeated, and multiple actions and practices related to diversity with assessment of the impact of these actions

Fourth order—attainment of transformative, culture-changing practices with campus connectivity and outreach to community and alumni as well as alignment of resources in support of these practices.

Juxtaposition of the mapping framework with the tertiary concepts of mobilization, implementation, and institutionalization suggests that fourth order change equates to institutionalization, while second and third order change are roughly equivalent to implementation phases. Higher-level implementation phases are differentiated by assessment of the impact of actions, an indicator of critical importance to organizational learning efforts.

Despite the seemingly linear nature of cultural change theories, the attainment of different levels of diversity culture change is not absolute and phases can overlap. For example, Adrienne Andrews, CDO at Weber State University, a large open access public university in Ogden, Utah, states, "Weber toggles between implementation and institutionalization."

Andrews, an African American female, describes the impact of university leadership on cultural change as follows:

> We have leadership from the top that is saying diversity and inclusion is critical for our ongoing success. It changed the conversation campuswide which also started to change the culture, which was one that wasn't actively exclusive but historically had the potential to be a lot better. People are now asking different questions about our climate. . . . Once people get here, how do we make sure that they feel included, that they have every opportunity to succeed, where does the scaffolding need to be? So different questions are being asked that make us have an improved climate for all people.

Furthermore, although the university has institutionalized programs such as an annual diversity summit in its eighteenth year and a Native American summit in its eleventh year, Andrews is working toward restoring a team-based college diversity structure that had existed a decade earlier with representatives from each college and division participating in a universitywide committee.

Given this overview of the phases of a diversity culture shift, we next identify specific characteristics of each phase of the change process.

Hallmarks of Mobilization

To begin the transformative process, a foundational approach for institutional leadership is to conceptualize a new organizational identity and integrate this identity in the institutional agenda through sensemaking.[48] The work of Karl Weick identifies seven key properties of sensemaking as fundamentally related to organizational identity construction: it is social; ongoing; retrospective and based on meaningful, lived experience; involves enactment or engagement within a given environment; takes place contextually based on cues; and invokes plausibility and coherence.[49] Leaders in this phase set direction and vision and give coherence to diversity initiatives through sensemaking and creating engagement.[50] As demonstrated in a study of six institutions making the most progress in a five-and-a-half year initiative of the American Council on Education's Project on Leadership and Institutional Transformation, sensemaking is a superordinate strategy that characterized the change process in all these institutions.[51]

Second, in the mobilization phase, an institutional diversity commitment is absent and the institution has not yet attained a common definitional framework leading to prioritized diversity goals. From this perspective, Allport's intergroup contact theory underscores the importance of having a common, superordinate goal to help individuals overcome bias, reduce in-group/out-group attributions, and develop more positive working relationships across difference.[52] This theory has particular relevance to the process of diversity change and overcoming resistance. In addition to a superordinate goal, Allport identified three other conditions necessary to reduce prejudice: equal status, intergroup cooperation, and support of relevant authorities.[53] Yekim, the Afro-Latino CDO in a private western university cited earlier, refers to this phase of diversity evolution as follows:

> The first layer is we are a culture that gets diversity but that do not get the work that is diversity. We understand the value and significance of diversity, but we don't understand what it takes to actually do the work in terms of material commitment and transformation. At some point even an effective synergetic framework is going to become ineffective in the growth of the work, because eventually it becomes an issue of time, of human band width, and not so much of getting it and doing a good job for that reason.

Hallmarks of Implementation

In the implementation phase, an institutional infrastructure for diversity has been put in place and both time and resources are allocated for diversity organizational learning. In order to promote the growth of diversity work, attention has been given to infrastructure and system design. Such organizational design includes recognition and rewards programs and material support for academic and administrative practices that foster diversity progress. The coordination of diversity organizational learning across various units including both academic, administrative, and student affairs is underway but not yet closely coordinated.

One of the principal challenges is bringing together all layers of the institution through the implementation phase due to differences in power and authority. For example, Marcy, a Latina female vice president of diversity at a private, Catholic urban university shares the difficulty of gaining active buy-in for diversity organizational learning from the deans and vice presidents:

> I worry about sustainability. The champion of the diversity learning initiative is the President who believes in it strongly. The organization itself loves it, because they are getting skills within their workplace to help them; they've invested in it. It's that next echelon at the university: below the president, the next executive level of deans and vice presidents. I have talked to them about it but I am not sure they understand. This year I am going to try to have a conversation with them at the President's retreat to build buy in. . . . Our deans are very decentralized and very powerful; they run their colleges and do their thing but they have got to get involved. That's going to be my challenge: to get to that next level at the university.

Diversity officers in our survey sample most often designated their campuses as in the implementation phase which typically include structural elements such as the creation of a diversity strategic plan, the creation of councils for diversity planning, the completion of a campus climate survey and establishment of accountability measures for addressing identified gaps, and the development of more comprehensive diversity professional development programs.

Hallmarks of Institutionalization

In this phase, institutional values have attained normative consensus and these values are reflected in behaviors, interactions, and concrete practices. Diversity programs are fully integrated into the organizational structure and have staying power.[54] With a shift in focus to outcomes and accountability, systematic diversity organizational learning programs transcend specific organizational silos and are connected throughout the campus infrastructure. Organizational design supports evolutionary diversity transformation. Concrete resources and material commitments to diversity are reflected in a sustainable diversity infrastructure. In sum, diversity leadership practices are integrally connected to institutional decision-making, strategy, goals, and measurable outcomes.

An example of organizational design that facilitates the institutionalization of diversity change can be seen at the University of Michigan and at Princeton University. Instead of budgetary oversight by the chief financial officer as is traditional at many institutions, the provost oversees the budget. This design places the academic mission at the heart of the budgetary allocation process. At the University of Michigan, the provost's office has the opportunity to align unit diversity goals with university mission during the annual budgetary process for the 49 administrative and academic units. As Robert Sellers, the university's CDO explains:

> So when I think about the [diversity] plan, I think the real strength and the real accomplishments that may not be evident at first lie in the fact that we have made it a part of the standard operating principles of the university. As a result, in the annual budgetary review process, each unit will have to report on progress in terms of the diversity goals that they have established. In these meetings, unit leaders will discuss why they have been successful and why they have not been successful and identify the resources or assistance that may be needed to move forward. Units have to report on their own compositional diversity, planned strategies, goals, and where they are in that process. That budget conversation is the one conversation at the university where every unit aligns their resources and efforts to their mission and is held accountable for their progress towards their goals. That's one thing that most people don't realize has a major impact in terms of change . . . and which gives me hope in terms of actual institutional change.

The institutionalization of faculty diversity programs at Columbia University is a significant example of the implementation of structural diversity practices with staying power. Since 2005, under the leadership of President Lee Bollinger, the university has invested $85 million in faculty diversity recruitment and retention initiatives that have impacted the recruitment of over 54 underrepresented and female faculty. The university provides resources that allow departments to supplement up to 50 percent of the hiring package over a period of up to three years for women and underrepresented candidates for full-time tenured or tenure-track faculty who are outstanding at the top of their field in the sciences or in other areas in which underrepresentation exists. The program is overseen by a senior faculty review board and involves a Request for Proposal (RFP) process in which departments submit specific proposals for funding to support faculty recruitment from underrepresented groups. Awards range from up to $250,000 each year for lab-based scientists to up to $125,000 for non-lab-based scientists. An award process also exists for the retention of underrepresented and female junior faculty for assistance with research and progress toward tenure. In October 2017, President Bollinger reinforced the university's ongoing commitment to diversity and inclusion by announcing plans to allocate $100 million over the next five years to recruitment and retention of diverse faculty.

To enhance university processes that impact structural diversity, the Vice Provost's Office in collaboration with the Provost's Advisory Council for the Enhancement of Faculty Diversity, has developed best practice guides for faculty mentoring and search processes and is now developing a guide for faculty retention. Other aspects of the institutionalized diversity portfolio include scholarly programming and discussions offered on a regular basis that promote diversity dialogue. According to Mitchell:

> We see diversity and inclusion discussed at all levels of the university and all search committee must use the best practice guides. . . . the combination of all of this has really begun to shift the culture. It's not perfect, definitely not. We have a long way to go, but we are very pleased with the progress in an era when we are seeing contraction.

Consider also the example of the attainment of normative consensus on diversity values that characterizes mature diversity progress. Christopher Washington, Provost at Franklin University, a private nonprofit university focused on adult learners, describes the common institutional mindset related to diversity:

> I feel that we are rather advanced on these ideas. I was just thinking about the major transformations our institution has gone through

> not just in serving diverse domestic populations but in really wiring ourselves as an institution to take on transnational projects. The number of international students who work on our campus, the number of international faculty and staff, and domestic diversity—African American, Asian, and Hispanic faculty who work on our campus and the diversity of our student body, and how the conditions at our institution enable diverse friendships. It just doesn't come up as an issue that needs to be solved with new mindsets.
>
> It is an approach I think that people have come to understand. If you are going to be a part of this university, you are going to have to respect diverse others and work with others who are different from themselves. . . . generally there is more of a climate of "we are all valuable to this institution" and what it means to achieve; whether you're a professor or someone who works in the bookstore, or deals with international students, or teaches in class part-time, or working in the student learning center helping people through tutoring: all of us contribute in so many meaningful ways to the mission of our institution and to support the needs of our learners.

As a reviewer for an accrediting body, Washington notes that other institutions have struggled with diversity and indicates that some contexts "do call for a transformative model that enables people to change their mindsets to deal with these artificial social boundaries that are created."

The examples of mobilization, implementation, and institutionalization of diversity referenced in this section offer a pathway toward the goal of diversity transformation through organizational learning and a sustainable cultural shift. Characteristics of mature models address compositional diversity, organizational design, resource allocation linked to diversity accountability, and the normative values of diversity in processes and interactions.

Forward-Looking Inclusive Excellence Change Strategies

Given the hallmarks of diversity transformation discussed in this chapter, we share examples of institutions that have undertaken systematic approaches to diversity learning through the IE change model. A leading-edge example of how to adapt the Inclusive Excellence (IE) change model to a specific institutional context is Missouri State

University's 2016–2021 Long-Range Plan. The plan identifies diversity and inclusion as one of six strategic areas and focuses specifically on creating a climate of inclusive excellence. The implementation of effective professional development that will increase cultural competence for faculty, staff, and students is one of the key goals in this area.[55]

The process of operationalizing the IE change model was initiated by a core group of stakeholders in 2008 and then became the central focus with the appointment of H. Wes Pratt as Chief Diversity Officer in 2016. Pratt, an African American male, explains the foundational process that will result in the development of an IE scorecard and performance indicators:

> We are trying to develop a diversity and inclusion action plan, and a part of that will be an Inclusive Excellence Strategic Implementation Plan where we can use the framework and four categories of Inclusive Excellence (access; success and equity; campus climate; learning development/curricular and co-curricular; and institutional commitment).
>
> . . . we know many of the things that we are going to be addressing, particularly in this next year, but our next step is to put together the framework of the IE change model in those four areas, establish what our priorities are going to be in each of those four categories of the framework, and then have an IE scorecard where we are able to actually track and monitor the success of the prioritized areas in those four categories. . . . part of the effort [was] we had to get everybody on the same page regarding what diversity is . . . even though some of us were operating with the AAC&U model and definitions . . .
>
> So with this long-range plan, which was just adopted by the Board of Governors, which is our policy-making entity for the university. . . . we actually have a section of the long-range plan that is called diversity and inclusion. We define diversity, inclusion, cultural competency and Inclusive Excellence. We created this baseline as the "ground floor" where everybody can be familiar with what we are talking about as we move forward to try to put together our framework for Inclusive Excellence at the university.

Another prominent example of an IE change model is the University of Delaware's plan titled, "Inclusive Excellence: An Action Plan for Diversity at the University of Delaware." Vice Provost for Diversity, Carol Henderson led an initiative to formulate the comprehensive plan which ties diversity practices to the quality and value of the educational process for all students.[56] The plan, described as a blueprint that is linked to the university's "diversity value proposition," is systematically organized. It begins with six principles for action that include an emphasis on creating opportunities for professional development in order "to leverage diversity as a core competency that has educational and societal value."[57] The plan is sub-divided into five sections, each with goals, accountability partners, and directed areas of improvement: 1) diversifying the academic community, 2) educating the campus regarding diversity goals, 3) improving diversity climate, 4) developing accountability for diversity goals, and 5) expanding engagement with the external community. The section on educating the campus includes the need to operationalize diversity learning, awareness, cultural competency, and professional development at all levels related to diversity and inclusion.[58]

Consider also the diversity learning initiative at the University of Wisconsin (UW) system of 25 institutions in addition to its flagship at Madison. The UW system has requested 2017–2019 biennial state budget funding to implement Fluent, a program requiring faculty, staff, and students to participate in training focused on cultural fluency. The program will be designed to enhance civility and respect and is one of five programs focused on improving the university experience with a collective cost of $6 million over two years.[59] The budget request was met with considerable skepticism by legislators such as State Senator Steve Nass, Republican, who described the proposal as "wasteful UW System spending on political correctness."[60] Nass described the proposal as "doubling down by suggesting the families of this state have failed in raising their children; that they are culturally incompetent."[61]

These examples demonstrate specific and concrete ways in which institutions have put in motion specific plans to address diversity and inclusion systematically across the varied, decentralized contours of the college or university landscape. We conclude the chapter with a case study of Princeton University that explores the ways in which an elite private research university is addressing historical and contemporary issues of inequality in reflective ways and moving forward purposefully on its diversity agenda.

Case Study III
The Progressive Course of Diversity Organizational Learning at Princeton University

This case study explores the ways in which administrative action and student activism have propelled significant advances in a diversity culture shift at a prestigious Ivy League institution. Princeton University is an elite private research institution located in Princeton, New Jersey, that has consistently reached the top of national and international college rankings in terms of academic quality, student experience and leadership. It is the fourth-oldest college in the United States and views its own identity as both a leading research university and a premier undergraduate college.[62] The 2017 U.S. News and World Report rankings found Princeton to be #1 in national universities as well as in best undergraduate teaching.[63]

For its first two centuries, Princeton's student body, faculty, and staff, were largely white and male and consisted of students from prosperous backgrounds. The university exercised discrimination against minority and Jewish applicants and there were few international students.[64] The first African American students were not admitted until 1942.[65] Princeton's efforts to create a more diverse learning environment began to take shape under the leadership of President Robert Goheen with the admission of women in the 1960s, the diversification of the eating clubs at the university, the hiring of the first African American full professor and first African American administrator, and the recruitment of minority students.[66]

The eating clubs at Princeton are privately owned mansions and are not officially affiliated with the university. The clubs have significant social impact determining not only dinner partners but in some cases affecting individual's "identity for life."[67] About three-quarters of its upperclassmen and women belong to one of the 11 dining clubs.[68] Six of the clubs are open to all students while five have entry criteria through an entry process called "bicker" with requirements that are not known to the university. Unlike single-sex Greek organizations, the clubs have been co-educational since 1990 due to a court decision.[69]

Princeton today has a diverse student body with an extremely large international cohort. The undergraduate student body includes 12 percent international students and 43 percent are American minorities.[70] Of the 7,998 undergraduate and graduate students, 41 percent are white, 16 percent Asian, 6 percent African American, 8 percent Hispanic, 3 percent two or more races, and 22 percent international.[71] Among its 773 tenured and tenure-track faculty, 77 percent are white, 9 percent Asian, 3 percent African American, 3 percent Hispanic, and 5 percent international.

While key elements present in this case study signify positive momentum in Princeton's journey toward inclusion, university administrators and students alike identify the need for further progress. Positive elements that characterize Princeton's intentionality in leading a diversity culture shift include the creation of a coordinated diversity infrastructure including approximately 40 positions with titles related to diversity, a flattened organizational design that enhances resource allocation aligned with institutional mission, a clear academic case for the value of a welcoming climate and diverse contributions, and involvement of governance structures in the change process including campuswide committees of faculty, staff, and student stakeholders.

As is true in other institutions of higher education, a number of contradictions complicate the picture for diversity such as external and internal pressures arising from the contrast between traditionalist views of alumni representative of a predominantly white elite and the perspectives of minoritized students regarding day-to-day experiences on campus as well as anger targeted against student activists by other students causing concerns among minoritized students for their own safety. The gap among constituencies is symptomatic of broad fissures throughout the higher education environment. Nonetheless, at Princeton, significant structural changes have occurred in resource allocation and staffing for diversity efforts, a systematic program for recruitment of historically underrepresented faculty, and organizational learning initiatives that support a diversity culture shift.

National Events and the Galvanizing Role of Student Activism

Princeton's progress on the diversity organizational learning continuum was spurred by the work of a Trustee Ad Hoc Committee on Diversity established under the leadership of now president emeritus Shirley Tilghman in 2013. The report generated by the committee presented the university's first comprehensive review of diversity involving multiple campus populations.[72] The Committee was co-chaired by Brent Henry, Vice Chair of the Board, and Deborah Prentice, chair of the department of psychology. Prentice was later appointed to dean of the faculty in 2014 and provost beginning in July 2017. Both Brent Henry and Deborah Prentice have played leading roles in diversity-related developments at the university.

The 2013 Trustee report specifically identified Princeton's "lack of progress" in diversifying its senior administration and faculty as well as its graduate students and postdoctoral fellows.[73] Further, it noted a significant demographic drop-off of women and minorities in the academic pipeline from students to senior faculty, with white Americans increasing 75 percent, while African Americans, Hispanic Americans, and Asian Americans decline in a range of 50 to 65 percent. Women also declined in the pipeline from nearly half of the undergraduate population to one-fifth of full professors.[74] The tenured/tenure-track pipeline at Princeton is dominated by Ph.D. recipients from elite institutions with 40 percent holding degrees from six universities: Princeton, Harvard, Yale, Stanford, the University of California at Berkeley, and the Massachusetts Institute of Technology (MIT). The report's recommendations include the proposal that each department create a strategic diversity plan through a phased program over three years and emphasized the need for active efforts to confront unconscious bias. It also called for the formation of a standing committee on diversity to monitor progress as well as strengthened trustee oversight of the work of senior administrators.

Deborah Prentice notes the significance of creating the university's first comprehensive statement on diversity and inclusion that was formulated in conjunction with the report. The fully featured Statement on Diversity and Inclusion describes Princeton's aspirations for diversity in terms of a global vision, the benefits of diversity in student intellectual and social development, and its commitments.[75] Prentice views the statement as an important milestone that enhances the university's ability to build on its diversity commitments.

In response to the Trustee report, President Christopher Eisgruber, who assumed Princeton's presidency in July 2013, recommended a comprehensive strategy to enhance diversity that included the development by each department of a multi-year diversity plan within three years. In November 2014, Provost David Lee began a phase of diversity investment and design involving several pilot programs. Deborah Prentice, Dean of the Faculty, expanded the Faculty Advisory Committee on Diversity and her office committed funding for a Target of Opportunity Program that would contribute the equivalent of 10 full-time faculty lines.[76] The program is directed toward the hiring of African Americans and Latinos in any discipline with an emphasis on hiring of underrepresented faculty in Economics, Engineering, and Business as well as Asian Americans in the humanities and some other disciplines. By 2017, the program already had met its objectives for increased hiring of underrepresented faculty. As Prentice remarks, "When you are paying attention, you are doing much better." The university also studied the career pipeline for diverse faculty and offered an array of formal and informal practices to support retention including mentoring and family-friendly initiatives.

With this foundational report as a backdrop, national political events drew further attention to the urgency of diversity culture change on the Princeton campus. Beginning in late 2014, student demonstrations at Princeton were sparked by the killing of unarmed African American men, Michael Brown in Ferguson, Missouri, and Eric Garner on Staten Island. Garner died after an illegal chokehold and chest compression by plainclothes officers, despite his repeated cries of "I can't breathe."[77] In both cases, the police officers were not indicted. The case study illustrates the dynamic impact of student protests linking events on the national stage with concerns about the need for structural diversity change on campus. Over a period of three weeks, a series of dramatic events related to diversity unfolded on the Princeton campus.

On December 2, 2014, a panel was held titled, "What kind of diversity: Is Princeton too narrowly focused on race and ethnicity rather than economic diversity?"[78] President Eisgruber participated along with Russell Nieli, a lecturer in the Politics Department, and two other panelists. Nieli had authored an article titled, "How diversity punishes Asians, poor whites, and lots of others" that explicitly connected the way that colleges talk about diversity as race and in particular, African Americans. In the article, he described diversity as "a politically engineered stew of different groups" driven by an ideology from the left and argued for the "enormous disadvantage incurred by lower-class whites."[79] The students reported that during the panel a question was raised, "Would you rather be poor and white or rich and black?" According to student activists, the panel served as a catalyst for their further engagement in pressing for diversity change.

On December 8, 2014 at an open meeting of the Council on the Princeton University Community (CPUC) convened by the president, an intensive discussion on diversity, equity, and inclusion took place during which a group of students expressed concerns about Princeton's climate. The students chronicled three primary categories of concern and shared these in a list with the committee: 1) transparency (particularly in resolving concerns about insensitive behaviors or comments); 2) awareness (relating particularly to curricula and a multicultural distribution requirement); and 3) resources for diversity-related programming.[80] In their words, "We have collectively identified our concerns related to current University policies that we believe implicitly promote a campus culture acceptant of racial animus and insensitivity."[81]

At the meeting President Eisgruber shared a statement on racial injustice and campus diversity that he had also posted that day on the university's homepage stating that "racial injustice has stained our republic from the moment of its inception."[82] He then charged the Executive Committee of the Campus of the Council of the Princeton University Community (CPUC) with developing recommendations on ways to improve the university's policies and practices relating to diversity, equity, and inclusion.[83] The CPUC Executive Committee subsequently appointed a Special Task Force on Diversity, Equity and Inclusion consisting of 51 members including faculty, administrators, staff, and students.[84] A Steering Committee chaired by Provost David Lee and three subcommittees were formed focusing on issues that roughly paralleled the list of student concerns presented to CPUC: Policy and Transparency; Academics and Awareness; and Structures and Support.

Meanwhile, on December 11, 2014, 200 students joined faculty and staff in a protest march of solidarity led by activists as part of a group called "Post-Ferguson at Princeton" which concluded with a 45-minute die-in simulating the four-and-a-half hours that Michael Brown's body was left on the street in Ferguson, Missouri.[85]

Work on the Special Task Force subcommittees began in earnest in February 2015. According to Genevieve, a minoritized student, the process was an awakening to her in terms of how universities work. After only four meetings of the sub-committee she worked on, a draft report of the Special Task Force was developed to coincide with CPUC's last meeting in May 2015:

> I think this was the first time I realized how universities and institutions work. We had four meetings to discuss all of

> diversity, equity and inclusion and somehow this is supposed to be comprehensive enough in some sort of form that students didn't help draft. . . . we only had four meetings: we couldn't discuss any one concept fully enough to issue something that people had really signed onto in the group. Our working group was talking about things like . . . cultural competency training for all faculty, diversifying the faculty, new programs around things like African American studies, Latina studies, each one of those is a huge subject and various ways that you could implement that.

Sean, a white male undergraduate, similarly expressed concerns about the findings being diluted due to the emphasis on many different aspects of diversity as well as the attenuated timeframes for discussion:

> Although it might have been initiated about talks specifically dealing with racial minorities and possibly socioeconomic status, it also became about gender, ageism, disability, and all sorts of different salient aspects of diversity and inclusion. That's right, but what it does sometimes is dilutes from the topics that were due to the taskforce. . . . There was a constant sense of tension that was unsettling. Because I remember at the third meeting, we had to send in all our comments to the steering committee . . . and I turned to my colleague and said "Where are our comments, everything is so diluted. Is there really a chance that we will get a final report back that has our stuff in it?"

Nonetheless, the report issued in May 2015 by CPUC offers substantive insights regarding campus climate based on data from an undergraduate senior survey and other qualitative reports. With considerable candor, the report notes the presence of "deeply embedded cultural norms" from society that may include overt racism, sexism, homophobia, and other types of biases. It further indicates that individuals with minority identities bear a disproportionate burden of negative experiences and that this burden is "fundamentally unfair and inequitable."[86] Minoritized students have encountered harassment and discrimination at times and been placed in situations that cause them to disrupt unproductive interactions based on their identities. The report clearly lays the groundwork for expectations for an inclusive campus climate that allows individuals to understand their own identities and also provides the content knowledge and skills to interrogate exclusionary social realities.[87]

The CPUC report provides a holistic overview of diversity experiences on campus and identifies six key areas for consideration with specific recommendations for improvement in each area. These areas include the student experience; addressing bias, discrimination, and harassment; academics and curricular offerings; and learning about diversity, equity, and inclusion outside the classroom. It particularly cites the distributed nature of staff members responsible for diversity and inclusion throughout the campus and the lack of consistent coordination among offices. The report recommends the creation of a plan to coordinate identity-based student resources, a strategy to address the issues faced by low-income and first-generation students, and the hiring of a senior administrator in Campus Life to oversee planning and coordination of diversity strategies. The subcommittee on Public Programming also recommended that the university host a series of public conversations in fall 2015 around themes such as building a culture of trust, free speech on campus, the imposter syndrome, and active inclusion.[88]

In creating a cohesive programming of diversity learning, the CPUC report describes the characteristics of effective professional development as follows:

> Effective training must be highly customized, based on the population, and provided in an effective sequence, in multiple forms. . . . It must be strongly encouraged and incentivized, and backed by key leaders. It must reach those who do not realize they need to learn and will not proactively seek training.[89]

Among its recommendations for enhancing diversity organizational learning, the report cites the availability of optional training programs and speakers or trainers who visit campus for single events, but critiques these offerings as lacking a consistent set of goals and being of uneven quality:

> The programs vary by semester and theme; there is no consistent set of trainings or goals and the quality is uneven. There are minimal opportunities for graduate students or faculty.[90]

The immediate commitments made by President Eisgruber after receiving the task force recommendations included the creation of the position of Dean for Diversity and Inclusion in the Office of the Vice President for Student Life. W. Rochelle Calhoun, an African American female, had been appointed as Vice President for Campus Life in 2015 and the new dean, LaTonya Buck, also an African American female, was hired in 2016. At Princeton, Campus Life oversees a substantial portfolio of six organizational units including Athletics, Campus Recreation, Office of the Dean of Undergraduate Students, Office of Religious Life, and University Health Services.[91] In addition, the Provost allocated over $400,000 in new funding to the Carl A. Fields Center as a "home base" for minority students, the Women's Center and the LGBT Center.[92] Furthermore, the Board of Trustees granted African American Studies academic department status in July 2015 and approved the concentration in this field.

In spring 2015 other events triggered concerns about the racial climate on campus. At a university-sponsored talent show event, an act titled "Urban Congo" featured a group of white members of the swimming and diving team who appeared bare chested in loincloths, drumming on garbage cans. Outraged by the lack of response from campus administration, student activists posted a video of the act on YouTube, leading to an outpouring of racial vitriol and naming the students who had participated in the demonstrations on Yik Yak, an anonymous social media application.

According to Achille Tenkiang, an African American senior who is pursuing an independent concentration in African Studies and Development:

> You walk around this campus, you sort of have this paranoia on your shoulders. These could be your classmates, your lab partners, your preceptors. . . . So I think for a lot of students of color and students from marginalized backgrounds, this was this sort of added weight that we carried on this campus during our sophomore year.

Achille describes how the student activists "sort of like put their bodies on the line in a very public way," adding that "there was a lot of backlash that these students were met with, not only from students but also from alums." He identifies the stress that minoritized students experience on a day-to-day basis and as a result, a number have requested time off to decompress:

> Students who take time off here, it's students of color, because of stress, there is an added weight, whether activists or not, oh wow, there goes like another one. This past week. . . . I have heard names of friends [who left to take time off].

Sean remarks that the backlash against student activists from other students and alumni led activists to be concerned about their safety:

> It was verbal assault . . . there were messages over emails. They would get messages from alumni targeting students on this campus, undergraduates, that were really just seething, and there was one time that there was a bomb threat. And so I guess to frame it, there was a bomb threat and students were fearing for their life and safety. It was really not crazy that someone could run in with a weapon. At the same time people were saying on social media, "stop complaining, stop whining."

Students became increasingly frustrated when no update was given on the actions taken on CPUC's recommendations for seven months. In the meantime, turnover occurred in top administration when two top administrators left their positions: the Dean of the College and the Vice President for Student Life.

On November 18, 2015, the Black Justice League, a student coalition, staged a 32-hour sit-in at President Christopher Eisgruber's office at Princeton and presented a petition with 1,000 signatures demanding that the university acknowledge President Woodrow Wilson's "racist legacy" and impact on campus culture and policy. The petition asked the university to rename the School of Public and International Relations as well as the residential college named after Wilson.[93] The student protesters also requested cultural competency training for faculty and staff as well as a general education requirement relating to the history of marginalized groups. In addition, they asked for the creation of a cultural space on campus for African American students.[94]

The sit-in of the president's office was not the first by minoritized students at Princeton. In 1996, a group of students occupied the office of then-president Harold Shapiro for 35 hours requesting greater representation for Asian Americans and Hispanics on the faculty and in the curriculum as well as the creation of ethnic studies programs.[95]

A day later, on November 19, 2015, President Eisgruber ended the standoff and along with Dean of the College, Jill Dolan, and Vice President for Student Life, Rochelle Calhoun, signed an agreement indicating steps that would be taken in response to the student demands. These steps included the commitment to request that the Board of Trustees begin conversations about Woodrow Wilson's legacy, consideration of the possibility of removing a mural of Wilson from a dining hall, designation of four rooms in the Carl A. Fields Center as Cultural Affinity Centers, and initiation of discussions about enhancing cultural competency training.[96] In addition, no disciplinary action would be taken against the students if they left the president's office peacefully.[97]

In response to the student demonstrations and occupation of the president's office, an opposition student group called the Princeton

Open Campus Coalition (POCC) was formed. The group indicated concerns about being labeled "racist" for those who questioned the protest and described the tactics of the demonstrators as coming "dangerously close" to intimidation. POCC also opposed efforts to remove Wilson's name from the School of Public and International Affairs as well as Wilson College. Further, the group expressed concern about required cultural competency training for faculty as threatening "to impose orthodoxies on issues about which people of good faith often disagree."[98]

Shortly afterward, the *New York Times* editorial board urged that in consideration of Woodrow Wilson's racist legacy that the board of trustees "not be bound by the forces of the status quo," indicating that "the overwhelming weight of the evidence argues for rescinding the honor that the university bestowed decades ago on an unrepentant racist." In response to the protests of the Black Justice League, the Princeton University Board of Trustees appointed a special 10-member committee chaired by Brent Henry, a '69 graduate of Princeton and Vice Chair of the Board. The committee was charged with reviewing the naming issues in light of Woodrow Wilson's role as university president in preventing the enrollment of African American students as well as the policies Wilson as president of the United States leading to the resegregation of the federal civil service.[99]

Resolution of the Woodrow Wilson Naming Controversy

By way of background, Woodrow Wilson, a professor of jurisprudence and department chair at Princeton, served as university president between 1902–1910 prior to his election as 28th president of the United States in 1913. During his time as Princeton's president, Wilson sought to make Princeton into a distinctive university and the best in the nation. His efforts included the reorganization of the faculty into departments and divisions, raising admissions standards, strengthening academic programs and expectations for faculty scholarship, and launching major fundraising efforts.[100] Nonetheless, Wilson actively prevented the enrollment of African

American students at Princeton.[101] The committee's report noted that Wilson had opposed admitting African American students to Princeton and stated that "the whole temper and tradition of the place are such that no negro has ever applied for admission."[102]

Despite his work in establishing the League of Nations for which he won the Nobel Peace Prize, as president of the United States, President Wilson oversaw the increased segregation of federal agencies including the Post Office and Treasury.[103] As president he approved measures for civil service employees that included demoting African Americans, installing curtains to separate African American and white clerical workers, and creating separate bathrooms.[104] In a meeting with a delegation of African Americans led by civil rights leader, William Monroe Trotter, on November 12, 1914, President Wilson told Trotter that he believed segregation in the federal agencies helped African Americans by preventing friction among white and African American employees. Trotter persisted indicating that separating workers based on race "must be a humiliation. It creates in the minds of others that . . . we are not their equals, that we are not their brothers, that we are so different that we cannot work at a desk beside them."[105] Woodrow Wilson grew angry over Trotter's tone and threw the delegation out of his office.[106]

After a process of information gathering that involved on-campus meetings, forums, discussions and the creation of a website that posted scholarly perspectives on Wilson's contributions and comments by 635 people, on April 2, 2016, the Trustee Committee on Woodrow Wilson's Legacy decided not to change the names of the Wilson School of Public and International Relations or the residential college. The committee noted Wilson's role as a visionary figure in international affairs and the changes he proposed in the residential colleges. The report issued by the committee urged consideration of historical context:

> Wilson, like other historical figures, leaves behind a complex legacy with both positive and negative repercussions, and that

> the use of his name implies no endorsement of views and actions that conflict with the values and aspirations of our times.[107]

The committee noted that Princeton has not been transparent about Wilson's views on race and indicated the need for the university to be more transparent regarding its historical legacy, especially as it pertains to race and to Wilson.[108] In October 2016, Princeton's unofficial motto coined by Woodrow Wilson, "Princeton in the nation's service" was modified to read "In the nation's service and the service of humanity." Supreme Court Justice Sonia Sotomayor, a Princeton graduate and former recipient of the university's Woodrow Wilson Award, expanded the informal motto as part of the university's evolution to serve the world at large.[109]

Reflecting on the outcome of campus demonstrations, the determination of the Trustee Committee on Woodrow Wilson's Legacy, and backlash from both alumni and student groups, Sean identifies the need for white students to have the tools to understand historical legacies of racism and how they are reflected in the campus culture:

> I think what was really troubling. . . . what was disheartening is you saw students whose citizenship or possession of this campus was constantly questioned, students of color by other students. You have alumni saying "Can't you be thankful? We give you donations where you can get the best education in the country. You shut up, you sit down and learn, and just be thankful." Whereas if there were even an instance if white students stood up or wealthy students or legacy students stood up there would be an immediate answer or they would be taken more legitimately. This disheartened me. The other thing we want to recognize is that the Woodrow Wilson conversation was incredibly nuanced, there were a million ways to approach it, and I think most students who were involved in those conversations did not have the tools to discuss what racism is, how it is in society what the legacies of such instances are and how they are reflected at Princeton.

Sean further describes the isolation minority students can feel on a daily basis due to the questioning of their concerns, but also emphasizes the substantive progress made by the administration in actions that advanced diversity progress within a relatively short period of time:

> As a white student on this campus, especially a white male student, it was hard to appreciate for students who would say, look at them; it was hard to understand how students of color are constantly reminded of their inferiority or questioned in a million different ways on this campus. . . . There was a serious measure of empathy that students weren't required to understand or gain, and that was one of the reasons that a lot of students pioneered an identity, race and ethnicity category for distribution requirements.

At the same time, Sean points out the ways in which the administration has worked more rapidly than in the past in terms of concrete actions and accomplishments:

> From someone who has worked with the administration since freshman year . . . I'll give credit to those administrators, Vice Provost Michele Minter, Cheri Burgess (Director for Institutional Equity and EEO) . . .
>
> . . . paradigm shift is not a thing here, even in issues that weren't around race and diversity. So what I saw was some of the demands were actually fulfilled pretty expediently. . . . Having an African American Studies Department put in place the summer after a year, having funding to completely redo and rename the Carl A. Fields Center and make that a new space . . . different hiring processes. . . . (etc.)

While acknowledging forward-looking steps taken by the administration, Achille Tenkiang reflects on the impact of alumni who experienced the campus differently than it was in previous decades:

> There are administrators who are dedicated, but I know that at the end of the day sometimes it does sort of feel as though they are beholden to other entities and parties. We do have a

> very strong alumni community. There's something to be said about the fact that every year almost 70 percent of the alumni donate. . . . It really feels like, yes, they did enjoy their time here but there also is a sense for them to sort of shape and sort of curate the Princeton experience and the Princeton community. And to some extent as undergraduates in this institution it kind of feels like we are undercut and our concerns are not given similar weight as concerns of alumni from thirty years ago who are not experiencing the way the world is now.

In support of this perspective, Genevieve underscores the contrast between traditional views of alumni about the university with the changing composition of the current student body:

> Princeton operates on the assumption that since it has been here since we became an actual country, it has worked for this long, it must be perfect. . . . I don't think people understand how a university that wasn't built for all the people who are here now could work for all the people who are here now.

While the controversy over the naming of the Woodrow Wilson School has attracted national attention, a number of initiatives are underway to allow Princeton to reexamine his legacy. Following the recommendations of the Trustee committee that the administration make substantive efforts to diversify campus art and iconography, a Campus Iconography Committee (CIC) was established coupled with an advisory group of faculty, administrators, staff, students, and alumni.[110] A Princeton Histories Fund was created to support projects that acknowledge the complexities of past legacies and reflect the university's diversity and inclusion such as through study of topics related to slavery, civil rights, and community activism.[111]

Structural and Compositional Diversity

Certain attributes of organizational design at Princeton pose both challenges and opportunities for diversity culture change. The provost is the chief budgetary officer for the university and responsible for coordinating the university's administrative and support functions with the institution's academic mission.[112] In this rather unusual structure, the academic department chairs report directly to the provost rather than an intermediate dean, except in the School of Engineering which has a senior academic dean. The dean of the faculty reports directly to the president as does the provost and the two senior administrators work collegially together. This flattened structural design can help break down the balkanization that arises due to the challenges posed by powerful deans.

Michele Minter, Vice Provost for Institutional Equity and Diversity, describes the benefits of a flattened organizational structure for diversity at Princeton:

> There are pluses and minuses, but in general we benefit from a more centralized structure. We have the budget and complexity of a great research university, and the culture of a smaller liberal arts college. That's a special mix that makes it possible to interact closely with colleagues and programs at all levels across the university. There can also be disadvantages: a strong dean of arts and sciences, for example, might efficiently corral a number of departments into a shared model. We often work with each department on a more individualized basis. . . . Overall our organizational model has allowed us to be well coordinated and aware of what is happening across the university.

Yet at the same time Minter points out the need to find a balance between institutional coherence and the potential for dynamism that can accompany more decentralized approaches:

> We are often advantaged by our centralized approach to problem solving and our ability to share information across the institution. At times, however, diversity can benefit from institutional dynamism. If you can test and promulgate lots of strategies because of a more decentralized structure, it can be an advantage. Any institution needs some of both. Given that we value being a coherent community, we try to make sure that the

> coherence doesn't become stifling and that we create enough space for creativity, innovation and heterogeneity.

Michele Minter views the flattened academic infrastructure at Princeton as a potential way to help overcome departmental silos. As she explains:

> Academic disciplines are very different from each other and administrative functions also differ. So it's inevitable that units have their own micro-cultures. We respect the way in which distinctive micro-cultures are a valuable part of academic disciplines but also try to avoid insularity and create opportunities for interaction and coordination. Our relatively flat organizational model means that there are fewer layers of process and hierarchy to navigate. In our environment, it is possible to maintain personal relationships.

Given the challenges of a decentralized structure, a number of comprehensive initiatives are underway that are designed to create a coordinated approach to diversity organizational learning across the university.

Self-Reflection on Inclusive Learning

In response to the recommendation of the 2013 Report of the Trustee Ad Hoc Committee on Diversity that all units formulate diversity plans, Human Resources (HR) began creating a planning framework for over 4,000 administrative and staff employees. Compared with many other institutions, the Office of HR has an unusually strong focus on diversity and inclusion with a staff contingent of five positions devoted to this area.[113] The Diversity & Inclusion unit includes a senior diversity and inclusion training specialist and a senior analyst in employee metrics. HR is headed by Vice President Lianne Sullivan-Crowley, a Cabinet officer, who reports through Executive Vice President Treby Williams. The HR Office closely coordinates its diversity work with Vice Provost Michele Minter.

Debbie Bazarsky, HR's Manager of Diversity and Inclusion who also served as Director of the LGBT Center for 14 years, sees the university's significant commitment to diversity as reflected in the significant number of positions across the university devoted to diversity. She explains:

> I feel like we are doing a lift institutionally. In my experience, when it comes to diversity and inclusion matters, most institutions of higher education put a Band-Aid on whatever is hemorrhaging. Princeton University is doing a lift, really investing the time, money, and resources to address diversity. HR is one part of the larger institutional diversity and inclusion initiative and strategy.

Substantial funds have been focused on staff training and shifting the culture with topics that included unconscious bias, microaggressions, and building diversity competencies.

In working with stakeholders over a one-year period, HR developed a consistent framework for diversity and inclusion plans with five pillars: recruitment and employee branding, retention (including performance management and professional development), leadership and accountability, learning and development, and building an inclusive culture. Each organizational unit has a committee for diversity planning, and HR met with the chairs of the committees to establish a common terminology and assisted units in developing a baseline, metrics, and goals for the areas of recruitment, retention, and climate. HR did not mandate what the diversity plans would look like but provided the five foundational pillars as a basis for mapping unit goals and objectives. Campus Life and University Services served as pilots for the diversity planning process and are now in year three of their plans. For example, University Services, an area roughly comparable to Auxiliary Services at many universities, has established a forward-looking diversity plan that establishes objectives for recruiting, performance management, training and competency building, climate and inclusive culture, and progress measurements.[114]

A second critical area of consideration reflected in the 2013 Report of the Trustee Ad Hoc Committee report was the diversity of senior administrators as well as their active role in diversity progress. Of particular interest is building leadership capacity to create and lead diversity teams effectively. The report emphasized that a commitment to diversity must begin at the top, and that cabinet members would be called upon to set expectations and assure accountability in their units. To assist units in understanding the current state of employee engagement, HR implemented a climate survey for senior administrators. The results were shared with Cabinet officers to discuss the top ten issues and provide an update on the survey findings.

The Division of Campus Life at Princeton has also launched a comprehensive array of initiatives to strengthen student engagement in diversity and community. W. Rochelle Calhoun, Vice President for Campus Life, recognizes the need to balance tradition and transformation. She views inclusion as a transformative process and "where the work is now:" Emphasizing the importance of "telling our story differently," she explains:

> Princeton's success in increasing the diversity on campus means that our work now is to become truly inclusive. Princeton is different now and to honor that difference we can't just tinker around on the edges. We need to think about and work on what it means to embrace all of the new aspects that are Princeton.

Specific initiatives under Calhoun's leadership that have strengthened student learning include the reorganization of the orientation program for new students and updated training of resident assistants. New undergraduate students now arrive on campus at the same time. The orientation program was reconceived with learning outcomes resulting from three identified goals: 1) promoting a smooth transition, 2) creating a sense of belonging, and 3) building a sense of community. Calhoun also has focused on training resident assistants in order to deepen their understanding of difference and strengthen their ability to work with students from diverse backgrounds. As Calhoun explains, "It's about fundamentally changing the way we do our work . . . and working hard to create a more integrated experience [for students]."

LaTanya Buck, the newly appointed Dean for Diversity and Inclusion in Student Life, is in the process of developing a framework of diversity competencies for student learning. The framework will be open for all to use and will create a model that connects with student learning outcomes related to diversity and inclusion. A pre-eminent goal for student affairs is, in her view, creating a sense of belonging for all students. Yet as is true at other campuses, student affairs are sometimes not viewed as an integral part of the student experience in relation to student learning, and relatively few opportunities exist for cross-collaborative work of faculty and staff in support of student learning outcomes. Buck would like to change that and asks, "How do we get the entire campus community committed to ensuring the success of all students?"

In her vision of student engagement, Buck emphasizes the importance of educating all students and her dream is that members of the campus community will see the cultural affinity centers "as not only support spaces, but spaces that educate and challenge every student here." With these aims in mind, she is working with a campus committee to assess and evaluate student affairs' current efforts, their relation to student learning outcomes regarding diversity and inclusion, and the language, mission, vision of Princeton and "who we say we are."

Buck says she feels Princeton has a genuine commitment to diversity. The level of resources devoted to diversity is, in her view, highly intentional and she indicates that the university "is on the cusp of change," which she embraces. At the same time, however, in light of the divisive external political climate, one of the current challenges is creating a sense of safety for underrepresented minority, LGBTQ students, international students, and Muslim students.

With these concerns in mind, following the divisive election of Donald Trump to the presidency on November 11, 2016, President Eisgruber issued a statement calling for Princeton to sustain its values and "steadfast commitment "embrace people of all ethnicities, religions, nationalities, genders and identities."[115] Given the increase in hate crimes on college campuses that followed the election, 300 faculty members signed a statement opposing racism and discrimination.[116]

Milestones and Aspirations

As seen in this case study, like other universities across the nation, student activists have played a significant role in pressuring for greater inclusion and progress in diversity learning such as through cultural competency training, campus iconography that reflects the diversity of the student body, and curricular change. Princeton has initiated self-reflective reports that examine historical legacies of exclusion but also point the way to continued and systemic diversity transformation. The 2013 Report of the Trustee Ad Hoc Committee on Diversity and Inclusion and the 2015 Report of the Special Task Force on Diversity, Equity and Inclusion offer explicit strategies that link the attainment of an inclusive campus environment to Princeton's success in the new millennium. While the learning environment has sometimes been contentious in terms of differing viewpoints such as divergent perspectives in the student body or contrasts with the expectations of more traditionalist alumni, the positive steps chronicled in this case study include leadership commitment; the development of an explicit statement on diversity and inclusion; attention by the board of trustees to the diversity of senior administration; introspective and candid self-reports of stakeholder committees; strategic diversity planning that includes the provost's office, human resources, and student life; forward-looking models of departmental diversity plans; significant resource and structural investments; curricular change; and enhancement of co-curricular diversity programs.

Notes

1. Leih, S., and Teece, D. (2016). *Campus leadership and the entrepreneurial university: A dynamic capabilities perspective*. Retrieved September 12, 2016, from http://papers.ssrn.com/sol3/Papers.cfm?abstract_id=2771269
2. Hamel, G. (2013). *Management 2.0: Revolutionary leadership*. Retrieved September 12, 2016, from https://breakfastbriefings.stanford.edu/briefings/management-20-revolutionary-leadership
3. Goodman, K. M., and Bowman, N. A. (2014). Making diversity work to improve college student learning. *New Directions for Student Services*, 147, 37–48.
4. Chun, E., and Evans, A. (2016). *Rethinking cultural competence in higher education: An ecological framework for student development* (ASHE Higher Education Report, Vol. 42, No. 4). San Francisco: Jossey-Bass.
5. Ibid. See also Kezar, A. (2008). Understanding leadership strategies for addressing the politics of diversity. *The Journal of Higher Education*, 79(4), 406–441.
6. Clayton-Pedersen, A., and Musil, C. M. (2005). Introduction to the series. In D. A. Williams, J. B. Berger, and S. A. McClendon (Eds.), *Toward a model of inclusive excellence and change in postsecondary institutions* (pp. iii–ix). Retrieved March 21, 2015, from www.aacu.org/inclusive_excellence/documents/williams_et_al.pdf
7. Williams, D. A., Berger, J. B., and McClendon, S. (2005). *Toward a model of inclusive excellence and change in postsecondary institutions*. Retrieved March 21, 2015, from www.aacu.org/inclusive_excellence/documents/williams_et_al.pdf
8. Eagly, A. H., and Chin, J. L. (2010). Diversity and leadership in a changing world. *American Psychologist*, 65(3), 216–224.
9. Williams, Berger, and McClendon. (2005). *Toward a model of inclusive excellence and change in postsecondary institutions*.
10. Ibid. See also Milem, J. F., Chang, M. J., and Antonio, A. L. (2005). *Making diversity work on campus: A research-based perspective*. Retrieved August 2, 2016, from http://siher.stanford.edu/AntonioMilemChang_makingdiversitywork.pdf
11. Feagin. (2006). *Systemic racism*. See also Chun and Evans. (2012). *Diverse administrators in peril*.
12. Kezar, A. J., and Eckel, P. D. (2002). The effect of institutional culture on change strategies in higher education: Universal principles or culturally responsive concepts? *The Journal of Higher Education*, 73(4), 435–460.
13. Kezar, A. (2014). *How colleges change: Understanding, leading, and enacting change*. New York: Routledge.
14. Schein, E. (2004). *Organizational culture and leadership*. San Francisco: Jossey-Bass.
15. Ibid., p. 11.
16. Ibid.
17. Bass, B. M., and Avolio, B. J. (1993). Transformational leadership and organizational culture. *Public Administration Quarterly*, 17(1), 112–121.
18. Schein. (2004). *Organizational culture and leadership*.
19. Kuh, G. D., and Hall, J. E. (1993). Using cultural perspectives in student affairs. In G. D. Kuh (Ed.), *Cultural perspectives in student affairs work* (pp. 1–20). Latham, MD: American College Personnel Association.
20. Kezar, A. J. (2007). Tools for a time and place: Phased leadership strategies to institutionalize a diversity agenda. *The Review of Higher Education*, 30(4), 413–439.
21. Kuh and Hall. (1993). Using cultural perspectives in student affairs.
22. Tierney, W. G. (1988). Organizational culture in higher education: Defining the essentials. *Journal of Higher Education*, 59(1), 2–21.

23. Rangasamy, J. (2004). Understanding institutional racism: Reflections from linguistic anthropology. In I. Law, D. Phillips, and L. Turney (Eds.), *Institutional racism in higher education* (pp. 27–34). Sterling, VA: Trentham Books.
    Tierney. (1988). Organizational culture in higher education.
24. Cohen, M. D., and Marsh, J. G. (1974). *Leadership and ambiguity: The American college president*. Hightstown, NJ: McGraw-Hill Book Company.
25. Feagin, J. R. (2006). *Systemic racism: A theory of oppression*. New York: Routledge.
26. Rangasamy. (2004). Understanding institutional racism.
27. Ibid.
28. Chun and Evans. (2009). *Bridging the diversity divide*. See also Hardiman, R., and Jackson, B. W. (1997). Conceptual foundation for social justice courses. In M. Adams, L. A. Bell, and P. Griffin (Eds.), *Teaching for diversity and social justice: A sourcebook* (pp. 16–29). New York: Routledge.
29. Feagin. (2006). *Systemic racism*.
30. Scharmer, O. (2009). *Ten propositions on transforming the current leadership development paradigm*. Retrieved September 14, 2016, from www.ottoscharmer.com/sites/default/files/2009_FieldBasedLeadDev.pdf
31. Brimhall-Vargas. (2012). The myth of institutionalizing diversity.
32. Milem, Chang, and Antonio. (2005). *Making diversity work on campus*.
33. Hurtado, S., Milem, J., Clayton-Pedersen, A., and Allen, W. (1999). *Enacting diverse learning environments: Improving the climate for racial/ethnic diversity in higher education* (ASHE-ERIC Higher Education Report, Vol. 26, No. 8). Washington, DC: George Washington University Graduate School of Education and Human Development.
34. Milem, Chang, and Antonio. (2005). *Making diversity work on campus*.
35. Museus, S. D., Ravello, J. N., and Vega, B. E. (2012). The campus racial culture: A critical race counterstory. In S. D. Museus and U. M. Jayakumar (Eds.), *Creating campus cultures: Fostering success among racially diverse student populations* (pp. 28–45). New York: Routledge.
36. Hurtado, S. (1992). The campus racial climate: Contexts of conflict. *Journal of Higher Education*, 63(5), 539–569; Milem, Chang, and Antonio. (2005). *Making diversity work on campus*; Museus, S. D., Ravello, J. N., and Vega, B. E. (2012). The campus racial culture: A critical race counterstory. In S. D. Museus and U. M. Jayakumar (Eds.), *Creating campus cultures: Fostering success among racially diverse student populations* (pp. 28–45). New York: Routledge.
37. Ibid.
38. Aguirre, A., Jr. (2000). *Women and minority faculty in the academic workplace*. San Francisco: Jossey-Bass.
39. Ackelsberg, M., Hart, J., Miller, N. J., Queeney, K., and Van Dyne, S. (2009). Faculty microclimate change at Smith College. In W. R. Brown-Glaude (Ed.), *Doing diversity in higher education: Faculty leaders share challenges and strategies* (pp. 83–102). New Brunswick, NJ: Rutgers University.
40. For fuller discussion of this phenomenon, see Chun and Evans. (2012). *Diverse administrators in peril*.
41. Curry, B. K. (1992). *Instituting enduring innovations: Achieving continuity of change in higher education*. Washington, DC: The George Washington University. See also Kezar. (2007). *Tools for a time and place*. Moses, Y. T. (2014). Diversity, excellence, and inclusion: Leadership for change in the twenty-first century United States. In D. G. Smith (Ed.), *Diversity and inclusion in Higher Education: Emerging perspectives on institutional transformation* (pp. 68–101). New York: Routledge.
42. Kezar. (2007). *Tools for a time and place*. See also Kezar, and Eckel. (2008). *Advancing diversity agendas on campus*.

43. Moses. (2014). Diversity, excellence, and inclusion.
44. Ibid., p. 68.
45. Halualani, R. T., Haiker, H., and Lancaster, C. (2010). Mapping diversity efforts as inquiry. *Journal of Higher Education Policy and Management*, 32(2), 127–136. See also Hurtado, S., and Halualani, R. (2014). Diversity assessment, accountability, and action: Going beyond the numbers. *Diversity and Democracy*, 17(4), Retrieved February 14, 2017, from www.aacu.org/diversitydemocracy/2014/fall/hurtado-halualani
46. Halualani, R. T., Haiker, H., and Morrison, J. H. T. A. (2013). *A comprehensive evaluation and benchmarking—2013*. Penn State. Retrieved February 14, 2017, from http://equity.psu.edu/workshop/assets/pdf/fall13/penn-state-comprehensive-evaluation-infographics
47. Ibid.
48. Kezar, A., and Eckel, P. (2002). Examining the institutional transformation process: The importance of sensemaking, interrelated strategies, and balance. *Research in Higher Education*, 43(3), 295–328.
49. Weick, K. E. (1995). *Sensemaking in organizations*. Thousand Oaks, CA: Sage Publications.
50. Kezar. (2007). *Tools for a time and place.*
51. Kezar and Eckel. (2002). Examining the institutional transformation process.
52. Allport, G. W. (1979). *The nature of prejudice* (25th ed.). New York: Perseus Books Publishing. See also Pettigrew, T. F., and Tropp, L. R. (2006). A meta-analytic test of intergroup contact theory. *Journal of Personality and Social Psychology*, 90(5), 751–783.
53. While a subsequent meta-analytic study using 713 independent samples in 515 independent studies found that the conditions Allport identified are not essential but facilitating factors, the study found that institutional support is a critical factor in producing positive intergroup contact results. See Pettigrew and Tropp. (2006). A meta-analytic test of intergroup contact theory.
54. Curry. (1992). *Instituting enduring innovations.*
55. Implementing the vision: 2016–21 long-range plan: Diversity and inclusion. (2013). Missouri State. Retrieved September 2, 2016, from www.missouristate.edu/longrangeplan/diversity=inclusion.htm
56. *Inclusive excellence: An action plan for diversity at UD.* (n.d.). Retrieved April 21, 2017, from https://sites.udel.edu/diversity/files/2017/01/Diversity_Action_Plan_PDF_R10-2jjcs1e.pdf
57. Ibid., p. 10.
58. Ibid.
59. Savidge, N. (2016). *On campus: UW-Madison rolling out cultural competency training for 1,000 new students.* Retrieved September 20, 2016, from http://host.madison.com/wsj/news/local/education/university/on-campus-uw-madison-rolling-out-cultural-competency-training-for/article_c3ea7957-fb65-568e-8d40-2e185d8677c0.html/
60. Ibid.
61. Nass, S. L. (2016). *UW system budget requires cultural sensitivity training for all Nass: Every employee and student will undergo cultural and conduct training.* Retrieved September 20, 2016, from www.thewheelerreport.com/wheeler_docs/files/0816nass.pdf
62. About Princeton University. (2017). *Inside Higher Ed.* Retrieved April 12, 2017, from https://careers.insidehighered.com/employer/1031/princeton-university/18/
63. Princeton University: Overview. (2017). *U.S. News & World Report.* Retrieved April 12, 2017, from www.usnews.com/best-colleges/princeton-university-2627
64. *Report of the trustee committee on Woodrow Wilson's legacy at Princeton.* (2016). Retrieved April 12, 2017, from www.princeton.edu/vpsec/trustees/Wilson-Committee-Report-Final.pdf
65. Armstrong, A. C. (2015). *African Americans and Princeton University.* Retrieved April 12, 2017, from https://blogs.princeton.edu/mudd/2015/05/african-americans-and-princeton-university/

66. Ibid. See also Martin, D. (2008, April 1). Robert F. Goheen, innovative Princeton president, is dead at 88. *The New York Times*. Retrieved April 12, 2017, from www.nytimes.com/2008/04/01/nyregion/01goheen.html
67. Yazigi, M. P. (1999, May 16). At ivy club, a trip back to elitism. *The New York Times*. Retrieved April 12, 2017, from www.nytimes.com/1999/05/16/style/at-ivy-club-a-trip-back-to-elitism.html
68. Canseco, S. (1999, March 18). Public and private: A look at Princeton and Yale's exclusive clubs. *The Harvard Crimson*. Retrieved April 17, 2017, from www.thecrimson.com/article/1999/3/18/public-and-private-a-look-at/
69. Yazigi. (1999). At ivy club, a trip back to elitism.
70. *Facts & figures*. (2017). Princeton University. Retrieved April 17, 2017, from www.princeton.edu/main/about/facts/
71. *Diversity data*. (2017). Princeton University. Retrieved April 16, 2017, from www.princeton.edu/provost/institutional-research/diversity-data/
72. *Report of the trustee ad hoc committee: On diversity*. (2013). Retrieved April 17, 2017, from www.princeton.edu/reports/2013/diversity/report/PU-report-on-diversity.pdf
73. Ibid.
74. Ibid.
75. A complete copy of the statement can be found in Appendix III of Report of the Trustee Ad Hoc Committee on Diversity (2013).
76. Patel, U. (2015). *New diversity initiatives include faculty hiring commitment, graduate recruiting programs*. Retrieved April 15, 2017, from www.princeton.edu/main/news/archive/S42/32/24O59/
77. Baker, A., Goodman, J. D., and Mueller, B. (2015, June 13). Beyond the chokehold: The path to Eric Garner's death. *The New York Times*. Retrieved April 15, 2017, from www.nytimes.com/2015/06/14/nyregion/eric-garner-police-chokehold-staten-island.html?_r=1
78. Mulvaney, N. (2014). *Princeton University president asks committee to review diversity, inclusion, equity policies*. Retrieved April 15, 2017, from www.nj.com/mercer/index.ssf/2014/12/princeton_university_president_asks_committee_to_review_diversity_inclusion_equity_policies.html
79. Nieli, R. K. (2010). *How diversity punishes Asians, poor whites and lots of others*. Retrieved April 14, 2017, from www.mindingthecampus.org/2010/07/how_diversity_punishes_asians/
80. *CPUC minutes*. (2017). Retrieved April 15, 2017, from www.princeton.edu/vpsec/cpuc/minutes/
81. Ibid.
82. Ibid.
83. Office of Communications. (2014). *President Eisgruber issues statement on racial injustice and campus diversity*. Retrieved April 13, 2017, from www.princeton.edu/main/news/archive/S41/80/69M77/
84. *Special task force on diversity, equity and inclusion*. (2017). Princeton University. Retrieved April 17, 2017, from www.princeton.edu/vpsec/cpuc/inclusion/
85. Liang, E. (2014). *Princeton students walk out, demonstrate against racism and police violence*. Retrieved April 17, 2017, from https://paw.princeton.edu/video/princeton-students-walk-out-demonstrate-against-racism-and-police-violence
86. *Report of the special task force on diversity, equity and inclusion*. (2015). Retrieved April 12, 2017, from www.princeton.edu/vpsec/cpuc/dei-report.pdf
87. Ibid.
88. Ibid.
89. Ibid., p. 12.
90. Ibid.

91. Day, D. (2015). *Calhoun appointed Princeton's vice president for campus life*. Retrieved April 17, 2017, from www.princeton.edu/main/news/archive/S43/68/71C67/index.xml?section=topstories
92. *Progress re the recommendations of the special task force on diversity, equity and inclusion*. (2016). Retrieved April 17, 2017, from http://inclusive.princeton.edu/sites/default/files/images/2016/Princeton_inclusion_2016.pdf
93. *#OccupyNassau meet black student's demands*. Retrieved April 17, 2017, from www.change.org/p/princeton-university-administration-occupynassau-meet-black-student-s-demands
94. Markovich, A. (2016, April 4). Princeton board votes to keep Woodrow Wilson's name on campus buildings. *The New York Times*. Retrieved April 17, 2017, from www.nytimes.com/2016/04/05/nyregion/princeton-board-votes-to-keep-woodrow-wilsons-name-on-campus-buildings.html
95. Sit-in at Princeton in president's office. (1995, April 20). *The New York Times*. Retrieved April 24, 2017, from www.nytimes.com/1995/04/21/nyregion/sit-in-at-princeton-in-president-s-office.html. See also Chong, J. (1995, April 24). Princeton students take over president's office. *Columbia Spectator*. Retrieved April 24, 2015, from http://spectatorarchive.library.columbia.edu/cgi-bin/columbia?a=d&d=cs19950424-01.2.26&e=--------en-20–32323--txt-txIN-Columbia------
96. Princeton students protesting Woodrow Wilson's legacy end sit-in. (2015, November 20). *The New York Times*. Retrieved April 16, 2017, from http://trocaire.ny.safecolleges.com/training/player/C-CRS-ITM-SUP_ROLE-TUT1/2B257922-1F84-11E7-A37C-4920CA20937E/

    Office of Communications. (2015). *University, students reach agreement on campus climate concerns*. Princeton University. Retrieved April 15, 2017, from www.princeton.edu/main/news/archive/S44/79/75E24/index.xml?section=topstories
97. Office of Communications. (2015). *University, students reach agreement on campus climate concerns*.
98. *Letter to President Eisgruber*. (2015, November 23). Retrieved April 17, 2017, from www.facebook.com/PrincetonOpenCampusCoalition/photos/a.1658902704349286.1073741828.1658896681016555/1658903194349237/?type=3&theater
99. The Trustees of Princeton University. (2016). *Report of the trustee committee on Woodrow Wilson's legacy at Princeton*. Retrieved April 14, 2017, from www.princeton.edu/vpsec/trustees/Wilson-Committee-Report-Final.pdf
100. Axtell, J. (2006). *The making of Princeton University: From Woodrow Wilson to the Present*. Princeton, NJ: Princeton University Press.
101. Whack, E. H. (2016). Princeton to keep Wilson's name despite his racist views. *The Associated Press*. Retrieved April 17, 2017, from http://bigstory.ap.org/article/7482d81f137d400e94e3c4b9783b9c00/racism-peace-prize-woodrow-wilsons-legacy-display
102. Ibid., p. 4.
103. Ibid.
104. Rothstein, R. (2016). *How Princeton's trustees failed history with their Woodrow Wilson decision*. Retrieved April 16, 2017, from www.alternet.org/education/how-princetons-trustees-failed-history-their-woodrow-wilson-decision
105. Lehr, D. (2015, November 27). The racist legacy of Woodrow Wilson. *The Atlantic*. Retrieved April 15, 2017, from www.theatlantic.com/politics/archive/2015/11/wilson-legacy-racism/417549/
106. Ibid.
107. Ibid., p. 12.
108. Ibid.

109. Dienst, K. (2016). *Princeton's informal motto recast to emphasize service to humanity.* Retrieved April 18, 2017, from www.princeton.edu/main/news/archive/S47/73/15K23/index.xml?section=featured
110. *Campus iconography.* (2017). Princeton University. Retrieved April 18, 2017, from http://evp.princeton.edu/committees-and-initiatives/campus-iconography
111. Aronson, E. (2017). *University projects will explore 'overlooked' topics in Princeton's history.* Retrieved April 18, 2017, from www.princeton.edu/main/news/archive/S48/47/62C02/index.xml?section=topstories
112. *Office of the Provost.* (2017). Princeton University. Retrieved April 18, 2017, from www.princeton.edu/provost/
113. *Staff directory: Human Resources.* (2017). Princeton University. Retrieved April 18, 2017, from www.princeton.edu/hr/progserv/contact/directory/
114. *A guide to the university services diversity and inclusion plan.* (n.d.). Retrieved April 18, 2017, from www.princeton.edu/uservices/us-together/DI_GUIDE_EN.pdf
115. *President Eisgruber '83 issues statement responding to the presidential election.* (2016). Retrieved April 18, 2017, from https://paw.princeton.edu/article/president-eisgruber-83-issues-statement-responding-presidential-election
116. Knapp, K. (2016). *Almost 300 faculty members at Princeton University sign statement opposing racism and discrimination.* Retrieved April 17, 2017, from https://planetprinceton.com/2016/11/17/almost-300-faculty-members-at-princeton-university-sign-statement-opposing-racism-and-discrimination/

4

REPRESENTATIVE APPROACHES TO DIVERSITY ORGANIZATIONAL LEARNING

> Nothing in science—nothing in life for that matter—makes sense without theory.
>
> Edmund O. Wilson, 1998, p. 56[1]

Why do diversity education programs fail to have transferable impact on campus culture, climate, and practices? Three principal reasons account for the failure of these programs.

First, diversity education often is not based on research or theory. A study of 178 articles on diversity training found that these programs tend to be atheoretical in both the way they are conducted and evaluated.[2] The literature supporting diversity training is similarly lacking in theoretical rigor. Over half of the articles (106) on this topic were simply descriptive without referencing any theoretical tradition or perspective.[3] The research literature itself reflects a reliance on nonexperimental methods as shown in another study with 77 percent of a total of 474 field studies on prejudice reduction interventions conducted without experimental means if evaluations were conducted at all.[4]

Second, diversity education programs typically are viewed as outside the mainstream of a university or college agenda. Viewed as "nice to-have" but not essential to institutional goals or purposes, diversity education programs frequently devolve into symbolic activities. When offered without clear criteria and objectives as well as a persuasive "academic case" for their relevance, diversity education programs will fail to attain measurable, sustained impact on prevailing culture, behaviors, norms, or attitudes.

Third, diversity programs often focus on the celebratory aspects of difference and do not address difficult sociohistorical realities or even touch upon the overt and subtle ways that exclusionary practices have been reproduced within institutional contexts. Research indicates that workplace diversity programs can threaten the status quo for whites due to their prominence in positions of power and concern about maintaining control as members of diverse groups gain greater influence.[5] Introduction of controversial programs such as antiracist training or discussion of systemic racism and historical structures of inequality can be unwelcome and cause considerable backlash.

Diversity officers face a number of conundrums in their efforts to provide diversity professional development across a campus ecosystem. They can be tasked with the schizophrenic role of providing diversity education to faculty, staff, administrators, and students in the face of limited campus interest, closet or active resistance, and the absence of leadership support for cultural change. They must straddle the fence between seemingly symbolic activity-oriented programs versus the need to hold difficult conversations on subjects that critique power and privilege. Like Goldilocks and the three bears, the porridge is usually viewed as "too hot" and "too cold," but never just right. With the limited tenure of presidential appointments, diversity officers must literally "roll with the punches" and may find sustained change difficult, if not impossible. Given their tenuous employment status, diversity officers often have less power than their titles suggest. As the research literature indicates, the relation between rank and power differs for minority group members who are less likely than majority group members to have the authority and power associated with the positions they hold.[6]

With these concerns in mind, in this chapter, we seek to identify common barriers and themes identified by campus diversity leaders in their efforts to facilitate an intentional and coordinated program of diversity organizational learning across a campus ecosystem. The focus of our inquiry is on diversity development programs offered outside the classroom, whether the participants in these programs are administrators, faculty, staff, and/or students.

While a number of institutions are funding diversity learning initiatives in light of student protests, the observations of our survey participants reveal that the "status" or financial resources of an institution are not necessarily determinants of the ability to deploy diversity organizational learning across a campus ecosystem. As a result, we explore the reasons why diversity education programs may or may not be successful and identify practices that enhance the potential for systematic diversity organizational learning.

Diversity Training Versus Diversity Education

At the outset, clarification of the difference between diversity training and diversity education seems particularly important. While the terms have often been used synonymously, a "research divide" between these two fields of inquiry has led to redundancy and a lack of integration.[7] Broadly conceived, diversity training tends to focus on management and staff while diversity education generally has been used to refer to curricular diversity learning for students.[8] Arguably, these two approaches would benefit from the research done in both fields and include contextual considerations, top-down and bottom-up perspectives, and create opportunities for demonstration and practice.[9]

In melding these divergent strands of the diversity research, the term "diversity education" has wider applicability as an umbrella term for diversity organizational learning programs. In this regard, Lisa Coleman, CDO at Harvard University, explains how traditional diversity "training" does not work, particularly in conjunction with faculty. Instead, she emphasizes the need for learning modules and platforms as the foundation for organization development:

> diversity training doesn't work. . . . We have moved away from training in the faculty area, training that has historically been considered to be typical training. And even in the HR and student areas, we have moved to more learning modules and

> platforms . . . as opposed to training. And so when we're thinking about the research, we have to think about what is the research that is the best at our institution, and the best for our institution is in two areas: implicit bias and working with organizational development.

Furthermore, due to Harvard's research mission, diversity organizational learning programs offered need to be research-based and field or area specific. As Coleman explains:

> For a university, particular at Harvard, when training is divorced from the research then it is going to be ineffective. . . . when I think about Inclusive Excellence, you have to be specific, you have to think about diversity and inclusion as belonging to a specific area and/or field. And then we have to think about an implementation strategy that works most effectively in those areas. . . . I also think Inclusive Excellence ultimately is about learning and creating a culture of learning versus a culture of experts. In the academic world we have a lot of experts, meaning faculty and core administrators who also need to understand the diversity and inclusion and belonging space.

Table 4.1 compares and contrasts key features of traditional diversity training and diversity education with a view to greater integration of these fields of inquiry. We have adapted the table to specifically address the needs posed by differing campus constituencies.

Table 4.1 Melding Diversity Training and Education Best Practices

Training Best Practices	*Education Best Practices*
Analyzes participant needs	Provides structured feedback frequently
Considers context and top-down impact	Uses performance metrics
Emphasizes skills and behaviors	Explores cognitive and affective processes
Employs demonstrations and practice (role plays)	Uses case studies, academic service learning as vehicles for experiential learning
Modified based on audience	Addresses disciplinary differences
Structured presentations coupled with exercises	Interactive discussions, study groups, team projects

Adapted from King, Gulick, and Avery, 2010, p. 897 (full reference found in note 7)

Transformative Learning for Diversity

Given the prevalence of atheoretical approaches to diversity learning, the purposes and goals of diversity training and education programs call for the examination of relevant learning theories. The multi-dimensional model of learning proposed by Kraiger and others addresses cognitive, affective, and behavioral aspects of learning.[10] Diversity training often aims at all three dimensions simultaneously without necessarily devoting attention to transfer of learning to the workplace or classroom.[11]

One of the most promising theories in relation to diversity training and education is the concept of transformative learning since it integrates a number of perspectives including the development of critical consciousness, individuation and reframing self-identity, a shift to a different developmental stage, relationship-building across difference, and a holistic learning process that integrates individual, spiritual, and social transformation.[12] Transformative learning is consonant with experiential forms of learning that address not only cognitive but affective dimensions and translate into modes of action.

Transformative learning, according to Jack Mezirow's original theory, is a process by which adult learners critically examine the assumptions that underpin values and perspectives and through reflection and rational dialogue transform these assumptions to be more inclusive and well justified.[13] An integrative theory of transformational learning also draws on the affective dimensions. Specifically, as Mezirow explains:[14]

> Transformative learning may be defined as learning that transforms problematic frames of reference to make them more inclusive, discriminating, reflective, open, and emotionally able to change.

Problematic frames of reference include democracy, citizenship, and justice, as well as diversity. Mezirow's theory builds on Jurgen Habermas' distinction between instrumental learning of knowledge and skills that improve performance and what he calls communicative learning or understanding.[15] Communicative learning necessarily requires translation into different ways of acting and interacting.[16]

A culturally responsive transformative learning model such as that developed by Charlyn Green Fareed, takes place within community-based learning groups through collaborative inquiry.[17] Such inquiry builds on four transformative learning goals: 1) creating culturally-sensitive learning environments, 2) fostering culturally inclusive learning experiences, 3) critically reflecting on and questioning culturally-shared meanings, and 4) group and personal learning through assessment methods that encourage freedom of expression.[18] While transformative learning can seem like a nebulous concept, it nonetheless addresses awareness of power and its relation to cultural differences[19] as well as realization of the common humanity that transcends difference. From this vantage point, it opens the door to changes in behaviors and actions.

Common Barriers to Diversity Organizational Learning

The main problem related to diversity learning is that relatively little research has been conducted on the effectiveness of diversity education programs. One of the early empirical studies that involved 785 human resource professionals across a variety of organizations found that half of the diversity training programs were judged as having a neutral or mixed effect while only one-third of the respondents believed the training to have been quite or extremely successful.[20] Over a decade later, a 2011 study of 1,500 senior managers in 50 organizations in private industry revealed that only one in four of the respondents indicated that the companies' learning and development programs were linked to the achievement of business outcomes.[21] The reason why such programs did not lead to organizational change was because the units that employees returned to after the training had not changed.[22] Since organizations are comprised of interactive systems, diversity progress cannot be expected to occur unless and until structures, processes, HR policies, and leadership direction support such change.[23] In other words, the target for change and development is not just the individual but the team, the department, and the organization.[24]

In this sense, higher education, like private industry, has tended to put the proverbial horse before the cart. Using the analogy of bricks and mortar, without the institutional bricks of leadership support, structural design, a supportive culture, and effective diversity policies, the mortar of diversity organizational learning will not adhere. Bearing these research findings in mind, alignment of diversity organizational learning programs with institutional mission, strategy, and policies strengthens the potential for follow-through. From a macro-perspective, diversity programs need also to develop desired organizational capabilities or collective abilities rather than simply focusing exclusively on individual competencies.[25] These capabilities are identified in an extensive body of HR empirical research studies to characterize well-managed organizations and include shared mindset, accountability, collaboration, strategic unity, and innovation among others.[26] Diversity is a differentiating organizational capability with outcomes for all stakeholders.[27]

Other environmental factors pertinent to the success of diversity learning in higher education include the institution's level or stage of diversity planning; its historical legacy; and the demographic makeup of the student body, faculty, and staff. These elements act in tandem as interlocking components of institutional systems that facilitate diversity learning. For example, although presidential leadership is essential in mobilizing diversity learning processes, when diversity organizational learning is viewed simply as an administrative responsibility, such learning will not garner the needed buy-in and ownership by the faculty and academic community.

Sustainable diversity organizational learning initiatives will benefit from patterns of action that avoid the following: 1) crisis-driven diversity agendas, 2) change initiatives mainly directed toward nondominant groups of students, 3) a focus on individual behavior change rather than organizational capabilities, and 4) locating change agents at the periphery of the institution.[28]

With these concerns in mind, we share the leading barriers that frequently impede the development of systematic diversity organizational learning:

1. *Lack of leadership commitment and a supportive institutional infrastructure to execute that commitment.* Researchers concur that top leadership support as evidenced through resources and commitment

as well as environmental context are critical factors in the effectiveness of diversity training programs.[29] Without a supportive context that enables the transfer of diversity education to the workplace and classroom, diversity education efforts will not lead to sustained change or discernable outcomes. Such diversity programs will have diversionary value without long-term results. For example, a study of 53 research assistants in a large university found that post-training implementation of skill-based diversity programs was affected by the perceived environmental consequences or the positive and negative reactions of supervisors and peers.[30]

2. *Siloed diversity learning programs.* Diversity organizational learning, like the higher education landscape itself, is often fragmented and confined to specific silos. Unlike compliance training which is generally highly structured and correlated with institutional policy and external mandates, diversity training can be episodic and activity based, without a clear connection to university policies or goal attainment. Without systematic coordination, diversity education often attracts only the same dedicated cohort of interested individuals. Diversity officers often struggle with limited audiences for diversity learning and the tendency for the same individuals to participate, a phenomenon known as "preaching to the choir."

   Different organizational units may offer training that is duplicative and even redundant. Take, for example, the situation described by Michael, an African American male diversity officer in one of the schools of a prominent private western university. Michael calls diversity learning at his elite institution "fractured":

> My institution is a confederacy: it has 9 schools. There are lots of things that are going on around diversity. There isn't a universitywide coordinated diversity program. . . . it's distributed. . . . there are lots of groups and some innovative things going on, but I can't say I have a handle on everything that is going on.

Michael asserts the need for leadership to step up what he calls "snails-paced" progress and establish accountability for outcomes:

> My issue is that we are all about these incremental projects . . . and little innovative efforts, but I am just not persuaded that diversity is one of the top priorities of the upper administration. And in my view, that's what has to happen if we're going to make more progress. It's got to be a top priority at the president's and the provost's level and I just don't believe it is. The thing is, if you talk to them, they are saying it [diversity] is important and they have certainly devoted some money to it. But if we are going to have more of this little, snails-paced progress, we have got to do something more audacious: we've got to start saying we are measuring people by the progress they make and the plans they put in place, and that's just not happening.

3. *Stand-alone, piecemeal programs versus sustained long-term efforts.* Often diversity education programs are offered as stand-alone workshops or seminars, without the benefit of an integrated approach or as part of a planned, systematic program.[31] In designing diversity programs, Susan Wilson, Vice Chancellor for Diversity and Inclusion at the University of Missouri at Kansas City (UMKC) describes the importance of a cohesive organization development framework using an inclusive excellence model tied to the strategic plan. Diversity learning programs, in her view, should be skill-based and linked to specific desired outcomes. Wilson, an African American female, explains:

> One of the things that I did when I came here is that I utilized an organizational development framework for diversity and inclusion efforts. I acknowledged the piecemeal approach causes diversity programs to fail. You have this program here and you have that little program there . . . and they are not attached to an organizational framework. And of course organizational learning is a part of that. We use the Inclusive Excellence model to guides strategies that we have implemented throughout the organization. Broadly speaking, we worked with all levels of our organization to engage in and implement our diversity strategic plan.

As Mario Browne, Director of Health Sciences Diversity at the University of Pittsburgh, observes, "those [diversity] initiatives that are of the one-and-done type are the least effective for long-term adoption or change." He also notes that "it is a common practice to do a diversity event, without any plan for followup and continued engagement." Browne indicates that having a high-level administrator, such as the Provost's office, partnering on initiatives and promoting them "adds a level of credibility."

4. *Limited resources and lack of collaboration among key administrative departments.* Limited resources are a significant issue in terms of the ability to offer university or collegewide diversity education. Jenny, an African American CDO at a public Midwestern research university, identifies the resource constraints she faces in providing universitywide diversity professional development as well as the territoriality arising from the perception her office may be intruding on other departments' responsibilities:

> I think one of the hindrances is resources. The other part is fear from some of the departments that we are stepping on their toes or that we are taking over someone else's job. For instance, if I say I want to do training for all new employees, HR might think that I am thinking that that's my job. But in actuality, I am just advising and working with HR in an advisory capacity in bringing the training for the staff. I think, that's been the pushback, we do this, stay in your lane. It's a territorial issue.

5. *Failure to include faculty in the development and facilitation of diversity education.* A repeated theme throughout the interviews with diversity officers is the need to have faculty integrally involved in both the development and delivery of diversity education. In this regard, Marybeth Gasman emphasizes the need for faculty to lead diversity seminars and the importance of speaking "the same language" as faculty. When administrators and the diversity officer engage in a kind of management or diversity speak, this approach will not resonate with faculty and the faculty will respond by tearing apart the methodology of the presenter. Gasman describes this linguistic disconnection as follows:

> They speak in a language that no one wants to listen to. . . . they speak in a student affairs language. They use terms like "a

> sense of belonging," instead of saying "belonging." I just think nobody wants to listen to that. Instead what they need to do is work with faculty and speak the same language as faculty and work on projects that way. . . . what faculty will do, if you say anything to them [in management or diversity speak], they will try to rip apart your methodology instead of focusing on the actual issue. That's what they do, it's a strategy. They're crafty. . . . I have seen them do it many times. That's how they get rid of candidates they don't like. That's how they get rid of student affairs people who come in and tell them something they don't want to hear.

6. *Absence of longitudinal assessment processes that track transfer of learning.* Few studies evaluate the causal effects of diversity interventions using social science methodology.[32] When comprehensive needs assessment and gap analysis are not conducted, a clear connection to desired learning outcomes cannot be made. For example, a survey of 108 diversity training providers found that while 95.4 percent identified changing behavior as a goal, the topics covered reflected a relative lack of focus on behavior.[33] Research indicates that assessment of the effectiveness of diversity training is the exception rather than the rule and few, if any, standards address the success or failure of such training.[34]
7. *Diversity fatigue and backlash.* One of the primary challenges faced by diversity leaders is how to frame diversity organizational learning efforts in the face of diversity fatigue and resistance. Several diversity officers in our survey emphasized that how difficult and controversial topics such as white privilege or antiracist training are introduced can affect the future viability of diversity organizational learning programs. In this regard, a study of 166 white male and female professionals and managers enrolled in executive MBA programs at four business schools found that the type of justification for workplace initiatives related to minorities influenced white backlash reactions. The degree of backlash was stronger in relation to affirmative action programs than diversity management programs.[35]

In addition, an analysis of data from more than 800 American firms revealed that companies can make things worse, not better, by programs that appear to be force-feeding employees or utilizing "command-and-control tactics."[36] To compound these difficulties, a review of nearly a thousand academic and nonacademic studies found that the positive effects of diversity training usually do not last for more than a day or two, and sometimes these programs can create problematic backlash.[37]

With these concerns in mind, Susan Wilson emphasizes that the process of diversity learning requires a degree of readiness and organizational preparedness:

> I think you always have to be mindful of organizational preparedness and what level they are in terms of their readiness to learn certain things. When I came into the organization, people were generally closed down to diversity. They either felt they had too much diversity training, or they thought that diversity had nothing to do with them. That's when we launched a 'Diversity Includes Me' campaign to send the message that all people are included when we say the words "diversity and inclusion."

A number of diversity officers in our study similarly cautioned about the approach to difficult topics. For example, Marcy, the Latina CDO at a Catholic university cited earlier, notes the importance of avoiding blame or creating guilt in the approach to difficult topics:

> When you have mixed audiences and you make people feel guilty and blame them, it is not conducive to learning. I have been to a few of those sessions. They were uncomfortable for me. You have to bring people along with you. . . . Shaming doesn't always work. And you have to get to that aha moment. . . . The nature of diversity officers is to bring people together; we all work together; we have to respect all cultures.

As can be seen from these testimonials, the approach to difficult diversity issues such as privilege or sociohistorical forms of exclusion necessarily avoids casting blame or guilt and requires creatively engaging individuals to reach a level of understanding through what Marcy calls an "aha" moment.

Creating channels for organizational learning can require diversity leaders to surmount both internal and external criticism. When such programs deal with substantive issues of diversity, difficult conversations can be anticipated. A report on leadership and race titled "How to Develop and Support Leadership that Contributes to Racial Justice" indicates that training programs that simply focus upon diversity practices, equal opportunity, and individualism, do not take into account how systems such as culture, institutional practices, and policies impact career and life opportunities for disadvantaged groups.[38] In the next section, we shall see specific examples of how diversity leaders approach the diversity education process to address underlying institutional realities and the need for change.

Strategic Diversity Learning Approaches

Given the common barriers that can thwart or even up-end diversity learning efforts, our focus shifts to creative strategies that capitalize upon existing resources, build buy-in, and create a sustainable and effective diversity learning platform. While among our survey sample there is significant variance in the level of support provided to diversity officers for professional development efforts, a number of common themes emerge. Campus readiness, alignment with institutional mission, recognition of underlying cultural norms and forces, and supportive leadership and institutional governance are the essential building blocks of a strategic approach to diversity learning. Following are recommendations of diversity officers for the creation of effective programs that lead to sustained diversity learning:

1. *Align diversity programming with institutional mission and goals.* Consider the perspective of Lisa Coleman, CDO at Harvard, who observes that when diversity learning approaches are not tied to the mission and the learning already occurring in the institution, these efforts will fail:

> It's why a lot of diversity and inclusion training fails. Because people think that if you go in for diversity and inclusion training at a macro level, you know three hours maybe three days, it fails. It is not part of the fabric of institution; it is not built in the mission; and it's not part of the learning that is already

> occurring in the organization. . . . you have to tie your diversity and inclusion and belonging to those learning modes and the mission, otherwise it'll be a failed enterprise.

This observation underscores the importance of clarifying the expectations for diversity programming and ensuring that diversity education is linked to institutional mission and goals as well as educational outcomes. Teresa, the CDO at an eastern religiously affiliated institution cited earlier, shares the view that diversity learning must be tied to a shared vision and institutional mission:

> A lot of it is around developing a shared vision, because I could talk until I am blue in the face about Inclusive Excellence, and again looking at the deeper culture, looking at how it is integrated across the institution, the organization. But if it's just me who's voicing that, then I think it is hard to move the whole campus. I think it is developing a shared vision: that's the key piece in terms of organizational learning. Then it's looking at campus systems, what do people need to know in order to do that, what do we need to know individually, what do we need to know collectively. That's the organizational learning piece.

Dennis Mitchell, Vice Provost for Faculty Diversity and Inclusion at Columbia University, shares how institutional alignment and support have created an environment that promotes ongoing diversity transformation:

> we are at a unique time in the university's history in that we have a wonderful alignment of vision in diversity and inclusion from our university president through our university provost through our board of trustees. That has really helped us to a place where we have been able to present universitywide initiatives on diversity going back for decades since 2005.

2. *Develop an iterative, strategic planning process for diversity education programs. Conduct regular assessment to identify training gaps and opportunities for improvement.* In developing strategic, evidence-based programs, a phase-based, iterative approach will facilitate the

attainment of long-term results and eventual institutionalization. These phases include pre-assessment, conducting needs analysis and gathering feedback, identifying gaps, launching and implementing the program, and evaluation. The frequent assessment and evaluation of diversity education programs will lead to subsequent modifications and revisions if programs do not facilitate the attainment of organizational goals. Benchmarking with peer institutions as well as outside industries is an important aspect of the evaluative process.

3. *Tie diversity learning efforts to change management theories germane to the higher education environment that allow appropriate sequencing of programmatic efforts.* As Christine Stanley, vice president and associate provost for diversity at Texas A&M University observes, CDOs must work effectively with the guardians of the culture and cultivating relationships with influential culture holders both inside and outside an institution.[39] The cultural change theories outlined in Chapter 3 provide a framework for operationalizing change management efforts.
4. *Ensure that diversity leaders across campus possess the needed competencies and skills for change management initiatives.* Since diversity is often viewed as a nebulous concept without a rigorous disciplinary basis, the attainment of diversity competencies by leadership can be overlooked. Nonetheless, a number of professional organizations have developed valuable competency models or benchmarks for diversity and inclusion practitioners.

For example, the Conference Board's model based on the input of 76 global diversity leaders provides an extensive evaluative matrix for diversity practitioner competence. Although primarily geared to the private sector, it has broad applicability to the issues faced by higher education diversity leaders. The close tie with the HR disciplines and change management are key features of this model. Seven primary competencies are identified and include sub-categories: 1) change management that encompasses organization development, communication, and critical interventions; 2) diversity, inclusion, and global perspectives that include cultural competence and negotiation/facilitation; 3) business acumen; 4) strategic external relations; 5) integrity; 6) visionary and strategic leadership; and 7) HR disciplines.[40]

In an effort to further professionalize the CDO role, the National Association of Diversity Officers in Higher Education (NADOHE) has developed standards of professional practice that identify the accepted characteristics of professional practice. The standards recognize the role of CDO as institutional change agent and require knowledge of the various domains that impact students, faculty, and staff as well as of the contexts and cultures that impact change efforts.[41]

5. *Develop collaborative initiatives.* Due to the limited availability of institutional resources, diversity officers frequently must exercise proxy agency to help secure desired outcomes by eliciting the skills and capabilities of a group working toward common goals.[42] Partnerships with allied offices including Human Resources and Affirmative Action provide the opportunity to provide compliance-related training or workplace-oriented programs. Our survey findings reveal that pooled resources allow budgetary costs for diversity education to be shared with the President's Office, the Provost's Office, and Human Resources. When institutions have placed affirmative action, Title IX, disability, and other compliance-related functions under the purview of the CDO, this broadened scope provides a greater range of budgetary support as well as the ability to consolidate programmatic offerings and resources.
6. *Pilot the change process.* A successful strategy described by several CDOs in our study is to pilot diversity action plans in one or several of the larger schools of a university. These plans are tailors to the diversity goals of the disciplines in the school or college and typically are reviewed for measures of progress and alignment with institutional objectives by the Chief Diversity Officer. Jenny, the CDO at a Midwestern public doctoral research university cited earlier, describes the review process of college plans and how academic program review and budgetary constraints have heightened the focus on outcome assessment:

> The next big step is really looking at what the colleges are proposing for their strategic plans, and part of their strategic plan must include having a plan on introducing or eliminating equity gaps for students of color. I don't think we have a model

> that we could say works: we have difficulty measuring that. I think we do a lot of different things and we are definitely not collaborating on them. But academic program review is forcing us to really at look at how we measure what's working, and what's not, and to shift resources to support what's working and eliminate things that are not.

Antonio, the CDO at an eastern research university cited previously, describes his strategy of piloting the development of diversity action plans in three colleges through the adaptation of a common instrument and indicators. The disciplinary-specific plans will be developed by college committees and reviewed by the appropriate academic governance bodies. They align with the university's diversity strategic plan that, in turn, is being formulated based on the outcomes of a campus climate survey:

> My intention with the three pilot projects is to demonstrate the validity of those practices in those three different colleges and then start disseminating that information. With that effort in conjunction with the diversity strategic plan based on the campus climate survey, we want to identify measurable goals for each of the different colleges that they can meet over time and develop trends for improvement. . . . So the whole idea behind the pilot project is to show the strength of these activities and then have each of the colleges replicate the effort. And I picked those colleges where I felt the deans would be the most sympathetic to these efforts.

Antonio is building buy-in through this piloting process that will allow him to make before a campuswide rollout.

7. *Use experiential forms of diversity education rather than lecture or didactic formats.* A purely didactic approach to diversity learning may be the least effective delivery mechanism since it creates further polarization and stimulates resistance and opposition. Experiential forms of learning, by contrast, are powerful ways of addressing emotional and affective responses to diversity. A number of universities are using improvisational theater to dramatize specific situations

and invite audience participation in the process. The use of theater depersonalizes situations and allows individuals to reflect on the responses of the protagonists in the theatrical enactments. At Temple University, for example, the InterACTion Theater Group created by the Learning and Development Department in Human Resources, provides in-service and external theater programs related to diversity issues.[43]

Ben Reese, CDO at Duke University, views standup training as the least effective form of diversity education. Instead he uses professional actors to dramatize specific situations in departments. As Reese explains:

> For the 20 years I have been here we have used professional actors. . . . My approach has been going to depts., seeing what their issues are with the director and then putting together 1 or 2 or 3 scenarios, often a compilation of several issues, but the scenarios are particularly related to their department. And then the workshop and the scenario plays out with professional actors with staff members engaging them.

Reese employs a number of practices to heighten experiential learning such as through programs that incorporate video clips to illustrate different issues and community-based initiatives that involve guided visits to local organizations and museums.

8. *Help constituencies/units take their work to the next level by focusing on what they value and do well and giving them specific tools that enable them to advance further.* Rather than a negative approach that emphasizes deficits, Robbin Chapman, Associate Provost and Academic Director of Diversity and Inclusion at Wellesley College, advocates the use of community mobilization techniques. By using Appreciative Inquiry as a learning approach, Chapman begins by "shining a light" on what organizational units or constituencies such as faculty and staff value and what they really care about. Appreciative Inquiry replaces a deficit-based approach with a strengths-based framework. Developed by David Cooprider and Suresh Srivasta at Case Western Reserve University in the 1980s, AI seeks to facilitate positive change

> and begins with what is already working well in an organization to build on successful practices and create a preferred future state.[44]

Chapman, an African American female and an electrical engineer by training, describes her process-oriented approach in three parts: 1) understanding "the irresistible need" that each constituency has, 2) building on what units already do well, and 3) providing tools that will enable areas or constituencies to take their work to the next level. As she explains:

> Some of the things that I use are community mobilization techniques. A specific framework that I use often is appreciative inquiry, and so basically that means that I solicit from whatever constituency it is . . . solicit things they think that they do well in what they do every day, things they are particularly proud of the way they do it. I get the list of all those things and without fail, there are a number of those things that help increase equity and the campus feeling welcome and inclusive. There are very few people that have absolutely nothing on that list. And what will happen is they won't be thinking about diversity and inclusion; they will just be thinking about how do I do my job or what am I proud of, what do I feel I do well; and so I identify things from that collection of items.

Chapman explains that "really making people see that what they do in their day to day work already contains important elements of what we are trying to get them to improve on, and to recognize and value and acknowledge that." This heightened awareness leads individuals and groups to understand that diversity and inclusion are not an "add on" but part of what they are already doing. She also emphasizes the importance of providing concrete tools to the groups she works with and practicing with those tools in the seminars and workshops she presents:

> The other thing that I think helps for all of those constituencies is I make sure that people don't just hopefully have an increased awareness of the issues or some of the concepts or the language or the framework, I make sure they have specific tools, that they have references to go out and find out and use on their own. In most cases I make sure that they practice with them during the workshop, enough to continue with them on their own. I brainstorm

> with them about what things they might be able to try next; what they feel empowered to do tomorrow or the next week. It's just really important that *they* articulate that, as opposed to me suggesting things, it just registers differently, and also they are going to articulate what they are currently empowered or at least have agency at least to try to do, and they have said it in front of a group of people.

In considering the readiness of different campus constituencies to discuss difficult issues, Chapman's approach is fine-tuned in terms of the frequency of opportunities she may have for interactions. For example, she has initiated year-long conversations with staff called "Community Conversations" that addresses issues such as intersectionality, ableism, and social class as well as an academic-year learning series titled, "Actualizing Equity" that has focused on the theme of privilege for the last two years.

With these best practices offered by diversity officers in mind, we consider now a differentiated approach to diversity learning for communicating with specific audiences that includes senior administrators, faculty, students, and staff. The value of such an approach is that it is tailored to the concerns and goals of each constituency and addresses the language, data, and issues that will facilitate greater engagement.

Cabinet-Level Interactions

In the model of diversity culture change that emanates from presidential leadership to the Cabinet level, engagement of the top leadership in the difficult issues around diversity is an important first step. A survey of over 771 diversity officers found that only one-third of the institutions responded that diversity education and training programs were directed to senior administrators.[45] Compared to programs directed toward faculty (43 percent), staff (51 percent), and students (50 percent), senior administrative programs have received the least institutional attention in terms of diversity training and education. This deficit is troubling, suggesting a disconnection between institutional rhetoric and organizational realities.[46] Or as Monique, an African American female CDO at an eastern Ivy League institution, puts it: "Leaders talking like talking heads" attend diversity events and "say the 'right things' but have no intention of really doing anything."

Larry Burnley, Vice President for Diversity and Inclusion at the University of Dayton cited earlier, describes how he initiated monthly Cabinet-level discussions titled "Courageous Conversations" at his former institution, Whitworth University (a private, Christian university in Spokane, Washington). First piloted by the president for one year, the program is now in its sixth year. Burnley indicates, "When I talk about this to colleagues they say it is unheard of, for that kind of sustained conversation to occur at the senior level." The program created a safe space for candid discussions based on scholarship relating to race, gender, sexual orientation, power and privilege, theological issues, and related topics.

The outcome of this program, in Burnley's words, was to move the campus paradigm at Whitworth University from one primarily focused on intercultural communication to one focused on diversity, equity, and inclusion. The Cabinet's focus, in turn, led to conversations among faculty, students, and staff as well as greater engagement of the Whitworth campus community:

> And so over time, I experienced the elevated importance of diversity, equity, and inclusion, because over time. . . . we moved from being a campus . . . whose work around diversity was being driven primarily by an intercultural communication framework or paradigm, towards one that was now also being shaped by the IE paradigm. . . . So there was a value. And so we created those spaces of conversations of engagement among faculty, students and staff.

Nonetheless, Burnley indicates that he still experienced pushback and resistance to the IE framework:

> There was still significant pushback, don't get me wrong. I am not claiming that there was total safety or lack of resistance for the work and position of the CDO and we had kind of arrived. We certainly had not. And there were faculty there who were making the same argument that we don't need this position, those who saw the work to advance IE as a threat to the mission itself: that this work around Inclusive Excellence and diversity could result in the moving the institution away from its Christ-centered mission

> to some form of social relativism. After a time, and as a historian of American education, I came to acknowledge the validity of this concern given the faith-based origins of higher education in the U.S. and the move toward secularization over time. At the end of the day, however, my work enjoyed strong support at the most senior level of administration and from other strategic positions from across the University. This was key in our ability to advance IE at Whitworth University.

Senior leadership discussions facilitate the success of diversity organizational learning by building overt commitment, sharing a common vocabulary, identifying key data points, ensuring alignment with institutional strategic plans and goals for improvement, and offering progress reports on specific initiatives underway.

Faculty Development for Diversity

Surprisingly, many faculty teaching and learning centers have been slow to incorporate diversity and multiculturalism as a critical component in faculty development.[47] A recent study reported the primary goals guiding faculty development programming among 164 faculty developers from campus teaching and learning centers. Diversity and multiculturalism were notably absent from the identified goals.[48] The reasons relate to three major factors: 1) a tendency to see diversity and inclusion as a matter of student development rather than teaching development, 2) reluctance to introduce diversity issues into the classroom due to a lack of training or the controversies that can arise, and 3) the risks involved in diversity training and the failure of past efforts due to uninformed or inadequate efforts.[49] Just as CDOs face the rise and fall of administrations, teaching and learning centers are not permanent fixtures with secure futures and the fortunes of such centers can come and go.[50]

Despite these trends, the role of teaching and learning centers is broadening to include organization development, despite the fact that the field may not have recognized this changing definition. The evolution of teaching and learning centers as change agents and leaders in addressing institutional changes is distinct from the traditional yet continuing emphasis on instructional development. The organization development aspect of faculty development addresses strategic institutional initiatives that include diversity and inclusion and the intersection with teaching and learning.[51]

In support of this expanded focus of teaching and learning centers, the greatest challenge in diversity organizational learning identified by diversity officers is the difficulty of engaging the instructional faculty. As Jeanne, an African American female CDO at a southern public research university observes:

> Generally, instructional faculty do not embrace the necessity or value of the training.
>
> There is instructional faculty resistance to the diversity-designated courses in the curriculum.

An important facet of diversity work with faculty is to provide empirical data and work with faculty as partners and allies in promoting diversity learning. As Norm Jones, CDO at Amherst College, explains:

> Because let's face it the faculty/administrator "divide" is a very real thing. People obviously expect the CDO to come in and have certain things to say and that's important. But I think when a faculty ally stands up and gives a real story about something that improved as a result of higher content knowledge, that goes a long way. . . . As I am thinking about workshops or even informal conversations in social settings . . ., that kind of thing is top of mind for me. Because yet another advantage of a small community, is that faculty can be really strong partners in terms of promoting the work. That's a big piece of it.
>
> I also think that it is offering resources. Faculty are faculty for a reason, right, they like work product. So the more data you have, the more context you have of how something should be framed, the more legitimacy you have in terms of socializing the value proposition with faculty. Some of whom may be reticent because they haven't seen compelling data about why a thing works. So it gets stuck in the ideology category for them.

Jones views faculty partnerships and advocates for diversity among the faculty as ways to facilitate faculty participation in diversity learning programs. If diversity programs are seen as adding value to faculty work through specific, practical strategies such as how to create an inclusive classroom or engage in culturally responsive pedagogical approaches, the likelihood of attracting faculty support is greater.

Administrator and Staff Professional Development

Professional development for administrators and staff is frequently provided in a wide variety of formats including modules within on-campus leadership development programs, stand-alone offerings, and campus-wide seminars offered on specific issues offered by external presenters. As Robbin Chapman points out, staff approach professional development from a different vantage point and with different motivations than faculty. In her view, staff see diversity education programs as "the opportunity to engage more deeply, to be a fuller citizen of the campus." Staff, she states, "are very eager to learn new things" and "figure out how to apply them" in order to become more active citizens of the campus. By contrast, since faculty "are in essence the fullest citizen of the campus," they will be attracted to programs that deal with inclusive teaching, mentoring, and interacting with students and resonate more with programs focused on leadership.

A leading example of an administrative program with impact on organizational culture is Harvard University's Administrative Fellowship Program. This program promotes leadership opportunities for members of historically underrepresented groups through a 12-month appointment in a mid- to senior-level paid position. Professionals are recruited from business, government, and education and the program, in turn, diversifies the Harvard and Cambridge communities.[52] The Administrative Fellowship Program offers on-the-job professional experience that helps build the leadership pipeline.

Consideration of diversity as an essential component of internal college and university leadership programs is less frequent than might be expected.[53] Iowa State's Emerging Leadership Academy, the Ohio State University Academic Leader Development Series, and the Stanford Leadership Academy all include a module that addresses diversity leadership issues.[54] Other types of diversity learning programs include diversity certificate programs for faculty and staff such as the University of California at Irvine's offering of ten seminars, the University of Pittsburgh's program of six workshops focusing on individual behaviors and policies, and Florida State's Diversity & Inclusion certificate that involves six classes and a Theory to Action project.[55]

Diversity officers and human resource leaders often partner to offer a wide variety of workforce diversity programs geared to general university audiences. An example is the cultural competency training for faculty and staff offered as a half-day seminar at the University of Kansas (KU). The campuswide half-day experience is now in its third year and in 2016 over 500 staff and faculty participated in the seminar, with 25 facilitators for each session. The conference covers issues of unconscious bias, micro-aggressions, critical social justice, and identity development.[56]

Student-Focused Diversity Programs

Student demonstrations have given rise to the development of required diversity training for incoming students such as the programs initiated at the University of Missouri at Columbia. At the University of Michigan, a pilot project was launched in fall 2016 with the goal of administering the Intercultural Development Inventory (IDI) or a similar instrument to all incoming students. The purpose was to assess the intercultural competencies, skillsets, and behaviors of students and to prepare a customized learning plan with followup assessments.[57] The results of the IDI will be used in standard, required, non-credit courses for all incoming freshmen that address perspective-taking, critical thinking, active listening, and other related skills.[58]

In a similar vein, following several racist incidents on campus, the University of Wisconsin at Madison has implemented a pilot cultural competency program called "Our Wisconsin" for 1,000 new students in fall 2016. With a cost of $150,000 to $200,000 in the initial year, the program will address topics that include identity, culture, and micro-aggressions and if successful will be slated for all incoming freshmen in 2017.

At Clark University, a private liberal arts college located in Worcester, Massachusetts, a presentation by the institution's new chief diversity officer introduced first-year students to the evolution of a new campus diversity vocabulary that includes micro-aggressions and micro-invalidations and described how to intervene when such subtle verbal, nonverbal, or environmental insults occur.[59] Like other diversity training programs that delve into more substantive issues, the program came under fire by some alumni, parents, and others. Clark President David P. Angel indicated that concern was expressed "that we may be overreaching in this area and that we aren't preparing our students for life's jagged edges."[60] He defended the training as designed to help students become aware of bias and discrimination, factors that preclude the free exchange of ideas. Emphasizing that the concept of micro-aggressions is supported in the research literature, President Angel further emphasized the need for discussion of difficult topics in order to reach common understanding from both a historical and contemporary standpoint.[61]

Oregon State University now requires students to take an online social justice course, and Virginia Tech similarly requires all first-year and graduate students to take an online course called DiversityEdu.[62] Following the video of fraternity members singing a racist chant that became public, the University of Oklahoma began a program of required diversity training for all new students called Diversity Experience Training based on the Intergroup Dialogue Model developed at the University of Michigan.[63] The University of Kansas also offers a course in Social Justice 101 for all new students as part of its Jayhawk Jumpstart program.

Developing a Taxonomy for Diversity Organizational Learning

Given our discussion of common barriers and themes in diversity education, how can campuses evaluate the state of diversity organizational learning? Gap analysis and diagnostic assessment are essential steps in laying the groundwork for diversity organizational learning. Consider the approach of Norm Jones, CDO at Amherst College, who focuses on ecological mapping as a diagnostic tool to evaluate how structures, processes, and practices of diversity and inclusion operate in the sometimes siloed operations of departments, units, and the college itself:

> When we think about the work of inclusion I really do believe that it is ecological work; so to the extent that organizations are predisposed to siloed operations, I think you can start with that data set and work your way in the other direction to ask why does it look that way over here and it doesn't look that way over there? And if you are asking those questions of the right stakeholders using appropriate data, it makes it very hard to take oneself off the hook for at least answering those kinds of diagnostic questions. I think at the end of the day even if it's for the sake of optics, most people want to be seen as willing participants in a collective enterprise as opposed to an outlier, siloed, myopic kind of subsystem that isn't actually helping to advance the strategic trajectory of the entire organization.

Jones begins the ecological mapping process by asking leaders specific process questions in order to uncover how diversity and inclusion is operationalized within their respective areas of oversight. He explains that he uses the institution's "existing scaffolding" to examine the diversity work underway rather than attempting to superimpose various plans:

> I try to use existing scaffolding to affirm and justify the work we are doing. I'm not coming in and creating a lot of new plans. I'm actually feeding back what the institution says about itself and what it wants as a way to frame the imperative.

The concept of ecological mapping provides a reference point for evaluation and benchmarking of diversity learning programs. In this regard, the diversity mapping framework introduced by Rona Halualani and others discussed in Chapter 3, represents an innovative strategy that allows institutions to depict graphically the progressive phases of diversity learning.[64] As a method of holistic inquiry the methodology gauges the endurance and sustainability of diversity work, with a recommended evaluation timeframe of five years.[65] From an ecological perspective, diversity mapping methodology offers a systematic process for evaluating the level and scope of diversity learning across the many dimensions of the campus landscape.[66] Using a visual coding approach, this methodology evaluates campus diversity initiatives across all divisions and units based on an electronic data analysis.

The Diversity Engagement and Learning Taxonomy Assessment (DELTA) analyzes diversity programs, curricula, and initiatives using a taxonomy that begins with knowledge and awareness, proceeds to skills and competencies, and invokes an understanding of power differences and the need for advocacy and social action to counteract injustice through innovative thinking at the highest levels.[67] The levels of the taxonomy are scaffolded, i.e. lower levels are included or enfolded within the learning process as institutions and individuals move to higher levels.[68]

A summary of the levels is as follows:[69]

Level 1—Appreciation, knowledge and awareness of diversity
Level 2—Skills and diversity competencies including the application of intercultural competence
Level 3—Interactive involvement including intercultural interaction
Level 4—Perspective-taking, personal investment, and recognition of self-other dynamic
Level 5—Critique of power differences, awareness of positionality, compassion toward others
Level 6—Social agency, action, and advocacy
Level 7—Innovation in problem solving using multiple perspectives, cultures, and contexts

This concrete methodology describes the attainment of outcomes that include intercultural competency, behavioral and attitudinal changes, and social agency in terms of self-other dynamics, compassion, and development of creative solutions to persistent social issues. Yet even this emerging methodology does not address the specific content of diversity-related programmatic offerings and events, the interrelationship of programs, and the resultant learning outcomes. Consequently, the sheer number of training activities, events, or programs is not necessarily a gauge of the relation of these activities to diversity goal attainment.

Key Questions in Evaluating Diversity Education Programs

In reviewing the array of approaches described at different institutions, a number of key questions can help guide the further development of diversity organizational learning. To begin, Dena Samuels proposes the following:[70]

1. Does the diversity education program bring a research-based focus into practice?
2. Are the trainings intersectional?
3. Are programs led by experienced diversity professionals?
4. Do programs offer ways to enhance self-awareness, skill building, and action planning around diversity?
5. Are diversity programs voluntary?
6. Are diversity education programs assessed as to their effectiveness such as in terms of the likelihood of more inclusive behaviors and practices by the participants?

To this list we add the following questions:

7. Are diversity education programs linked to strategic diversity goals and the institution's mission/vision?
8. Is diversity program planning iterative, i.e. based upon a cycle of needs assessment, benchmarking, planning, evaluation, and feedback?
9. Are diversity programs tailored to the needs of specific audiences?
10. Do programs address sociohistorical contexts and the workplace implications for equity, diversity, and inclusion?

As Samuels proposes, if the answer to any of these questions is "no," then the final question is this: are the investments of funding and time devoted to diversity education worthwhile?

A central theme of this chapter is the need for research-based, data-driven diversity learning approaches that address the needs of faculty, administrator, staff, and student stakeholders. Differentiated strategies that address the needs of constituent groups in terms of language, content, modality, and expected outcomes are critical to the success of diversity organizational learning programs. An iterative process includes specific, ongoing evaluation that begins with needs assessment, provides specific offerings related to identified needs, and addresses the effectiveness, transferability, and outcomes of diversity education. Our interviewees emphasize the importance of gauging organizational readiness, understanding cultural tensions and underlying resistance, and building alliances with diversity advocates particularly among tenured faculty. Other promising approaches include piloting diversity learning programs in the largest schools or academic units, creating a positive framework that builds on the strengths of stakeholders' existing work and vision of a desired future, and developing experiential learning programs that build on the theoretical foundation of the inclusive excellence change model. We have also seen specific ways in which diversity officers have tackled difficult topics and built buy-in among different institutional audiences.

The diversity learning strategies described in this chapter will help create a strengths-based cultural change framework calibrated to the needs of different constituencies and linked to institutional mission. With these insights in mind, we conclude the chapter with two case studies: the first at Lehigh University addresses the prominent role of diversity in academic transformation, while the second at the University of Maryland at College Park examines the evolutionary movement from compositional diversity to a focus on inclusive diversity learning programs.

Case Study IV
Academic Prominence and Diversity Culture Change at Lehigh University

This case study explores the ongoing process of fostering diversity culture change at a predominantly white, private research university. Lehigh University is an academically competitive institution located in Bethlehem, Pennsylvania, on the outskirts of New York City. It is ranked among the top 50 national universities in terms of the full range of degree programs and faculty research and among the top 25 in terms of most innovative schools.[71]

The case study focuses on the steps taken by both administrative, faculty, and student leaders to build an academic infrastructure and inclusive campus climate that promotes diversity learning. One of the more notable areas in which Lehigh is differentiating the academic landscape for diversity is through interdisciplinary learning. The university has articulated the goal of interdisciplinary learning across all undergraduate programs in order to create an intentional pathway for students to realize the educational benefits of diversity.

Lehigh was established in 1865 by railroad entrepreneur, Asa Packer, as an all-male institution with a focus on mathematics and science education. Although the first women were admitted as graduate students in 1918, the decision to admit female undergraduates was not made until 1970. The first woman associate professor was hired in 1965.[72]

The university currently is comprised of four colleges: Arts and Sciences, Business and Economics, Education, and Engineering and Applied Science. In recent years, the College of Arts and Sciences has grown in enrollment and academic stature with 32 percent of the undergraduates compared to 25 percent in the College of Business and Economics and 31 percent in the College of Engineering and Applied Sciences.[73] Despite the growth of the College of Arts and Sciences, the university's historical focus has been on engineering and business administration. This emphasis has influenced the culture at Lehigh University. As James Peterson, Director of Africana Studies and Associate Professor of English, explains:

> Lehigh is known as an engineering school, more recently it has been identified as one of the leading colleges of business and economics in the country and that identity is very strong. And because Lehigh is a very traditional institution it holds onto its traditions very tightly.

Like many other institutions of higher education, Lehigh's progressive journey reflects the tension between historical tradition and change. Given the social forces of inequality present in American society as a whole, the diversity efforts underway at Lehigh reflect intentional actions to create a welcoming and inclusive environment for learning.

The Impact of Leadership Changes on Diversity Progress

New presidential leadership has focused on enhancing Lehigh's academic prominence, linking such prominence with an inclusive and globally aware environment for student learning. The university's "Principles of Equitable Community" refers to the development of future leaders in a global society as an essential educational commitment.[74] The statement of principles also affirms the "inherent dignity" of all members of the campus community and the importance of an equitable and inclusive community.

The appointment of a seasoned academic administrator, John D. Simon, as president on July 1, 2015, signaled a sea change in diversity leadership. Simon, a white male, previously served as executive vice president and provost at the University of Virginia where he founded the global studies major and opened an Asian campus. In this prior role, Simon sought to break down the separate "silos of command" among schools and departments and increase the relevance of the Provost's office.[75]

During the leadership crisis at UVA when the Board of Trustees forced the resignation of President Teresa Sullivan, Simon's courageous stance and alliance with a unified faculty helped turn the tables in Sullivan's favor. Risking his position and the ire of the Board, Simon delivered a seven-minute speech to the Faculty Senate:[76]

> I find myself at a moment when the future of the university is at risk and what our political leadership value in the university is no longer clear.

> The board actions over the next few days will inform me as to whether the University of Virginia remains the type of institution I am willing to dedicate my efforts to help lead.

In fall 2016, Simon introduced an ambitious ten-year plan for Lehigh's future growth, "The Path to Prominence." The creation of the plan signaled important changes for diversity, specifically the following:

1. Attracting a more diverse student body from a wider geographic base;
2. Increasing faculty, staff, and student diversity;
3. Enhancing an interdisciplinary approach to undergraduate education with an emphasis on outcomes and a global mindset and experience. Interdisciplinary perspectives will be part of every major, minor, and certificate program.[77]

Over the next seven years the university intends to increase the faculty by 100 new members, add 1,000 undergraduate and 500 to 800 graduate students, and establish a college of health among other initiatives. President Simon further explained the importance of a liberal education and the link between student learning and future careers in a global society:

> Students need an education that prepares them for success in professional and civic life. The world needs citizens who possess the creativity, civic learning, communication skills, and critical thinking that the arts and humanities provide, so they can take on the task of solving an increasingly complex set of challenges.[78]

In supporting the creation of a "more dynamic institutional environment" through the Path to Prominence plan, Provost Patrick Farrell, a white male, acknowledged the potential for increased discomfort but linked the greater focus on diversity with a liberal education that would prepare students for future careers in a global society.

President Simon also announced the creation of a new Cabinet-level position, Vice President for Equity and Community. Donald Outing was subsequently appointed to serve as the inaugural Chief Diversity Officer reporting directly to Simon. As founding director for the Center for Leadership and Diversity at West Point, Outing, an African American male, had spearheaded STEM (science, technology, engineering, and math) outreach strategies for middle school students. The creation of this universitywide position previously had been recommended by the Council for Equity and Community, but the former Lehigh president instead created a vice provost position for academic diversity, a position now held by Henry Odi.

As a touchstone in its diversity journey, Lehigh's eloquent statement "The Principles of Our Equitable Community," articulates the university's commitment to academic freedom, open expression, a culture of unity and respect, and the need to confront and reject discrimination.[79] It has been endorsed by senior administration, university faculty, undergraduate and graduate student senates, staff, Board of Trustees, and the Alumni Association Board of Directors.[80] Following the divisive 2016 presidential election, President Simon, Provost Farrell, and Patricia Johnson, Vice President for Finance and Administration, issued a message to the campus community stating, "These Principles are not just words on paper; they are what we believe in and what we expect of our entire campus community."[81]

Despite such high-level endorsements, a clear challenge remains in terms of building support for the principles within the diffuse and decentralized domains of the university infrastructure. Monique, a minority female department chair, indicates the need to translate the principles into action through structural changes in organizational processes such as promotion and tenure and teaching evaluations. As she explains:

> I think there's acceptance perhaps at the level of principle, but I'm not sure about the actual engagement in action. I think there are some segments of the community that have embraced

> it and have always been aligned with these principles and teach to the principles. . . . however, people are still kind of wondering exactly how to include this in the classroom and the curriculum.
>
> We need a lot more training about these issues in terms of how you engage in difficult dialogues; I know there have been workshops offered . . . that address some of this . . . and they have talked about inclusive teaching and classrooms, but again oftentimes it is to the group of faculty that are already engaged and committed to this. I am not sure how largely this is being adopted. I think part of it is about what kind of motivations and incentives are in place. It is unfortunate that I have to say that. But I think, for example, in merit raises or in promotion and tenure documents or in teaching evaluations, there is really nothing related to issues of diversity; so structurally there are some barriers that don't allow it to be fully embraced.

A program of organizational learning initiatives launched by the vice provost for academic diversity is now underway to operationalize the principles.

In addition, while Lehigh's leadership and strategic planning goals are consonant with the tenets of Inclusive Excellence, the university still has some distance to travel in building a more diverse administration, faculty, staff, and student body. The demographic makeup of the student body of 5,080 undergraduates and nearly 2,000 graduate students is 65 percent white, 9 percent Hispanic, 8 percent Asian, 4 percent African American, and 9 percent non-resident aliens.[82]

Compared to its academic counterparts, the small number of African American students at Lehigh is a matter of specific concern. For example at Amherst College, 12 percent of the student body is African American and 44 percent of its students from the United States are minorities.[83] Of the 540 full-time professors at Lehigh, 69 percent are white, 14 percent Asian American, 3.5 percent African American, 3.3 percent Hispanic, and 4.6 percent are international faculty.[84] The larger percentage of Asian American professors is distributed across three of the university's four colleges and 5 of its 31 department chairs or 16 percent are Asian American, a representation atypical for many institutions.[85] Only one department chair is African American and none are from the Latino\a racial-ethnic group.

The Aftermath of Incidents of Racial Harassment

Several years ago, the campus experienced a number of incidents of racial harassment, resulting in a voluntary resolution agreement with the Office of Civil Rights (OCR) reached in 2014. The complaint filed by a 1977 alumna, Susan Magaziner, alleged the university's failure to provide appropriate responses to incidents of racial harassment.[86] Press reports indicate that Ms. Magaziner alleged that the university had not reported the vandalism of a multicultural student residence, Umoja House, on November 6, 2013, as a hate crime when the residence was spray painted with the "N" word and egged.[87] Magaziner stated she viewed the Umoja House incident as "the final straw," referring to previous incidents and calling for action in response to complaints of discrimination.[88] President Alice Gast issued an immediate statement describing the egging as a "hateful act" and called for the university community to stand together "against all acts that are motivated by intolerance and aggression."[89]

Specific outcomes included in the voluntary resolution settlement were these: development of an anti-discrimination statement, staff and student training on racial harassment, a racial harassment policy, and remedies for racial harassment through student processes. The voluntary OCR resolution agreement also included a commitment to conduct an annual campus climate survey ("The Lehigh Survey") and share the results with OCR.[90] Additional provisions of the agreement addressed revision of the University's harassment policy, issuing of an anti-discrimination statement to all students and providing education and training on racial harassment as well as required online training.[91]

In September 2013 another racial incident at Lehigh involved a fight at a local restaurant and bar between football players and members of the Phi Kappa Theta fraternity over the allegation that a racially derogatory term was used.[92] An African American Lehigh football player was arrested and sentenced to one to two months in jail for simple assault and drunkenness.[93] Leah, a minoritized student, describes the incident:

> There was a fight that broke out after hours at a bar between black men and white men—Lehigh students. And in the next couple of days the Brown and White newspaper as well as the local newspaper came out with a story on that fight . . . but only put pictures of the black students. There were no pictures of the white students even though it was a fight that involved equal numbers on both sides. Some of the black students were football players. So they had multiple pictures to choose from, but they chose random pictures from the social media pages that made them look like any other media that wants to make black students look like monsters. At this time we asked the university to respond to this by protecting the black students or if they were going to shame the report and the entire thing on the black students and disclose the pictures and names of the white students. We never got any response.

The lack of responsiveness of the administration to this incident as well as the egging of the Umoja House led a group of eight students to organize a group called "From Beneath the Rug" (FBR). The group focused on fighting for the rights of minoritized students and addressing issues they believed had been swept under the rug. A protest took place on October 4, 2013, when students walked through the campus wearing black and white formal clothing with duct tape over their mouths.[94] In Leah's view, the protest was designed "to draw attention to the U house incident and to show that university did not care about diversity but only used diversity as a marketing tool."

The university took a number of measures to address equity and inclusion at Lehigh following these incidents. President Gast announced that she was reenergizing the Council for Equity and Community, meeting regularly with members of FBR, and was creating a permanent student group that would focus on campus climate issues.[95]

In 2015, during the national protests led by the Black Lives Matter movement, some African American students felt unsafe and feared for the safety of their families. The students requested a meeting with Henry Odi, Vice Provost for Academic Diversity, an African American male and Lehigh doctoral recipient with significant university service, and Provost Farrell. At the meeting that was quickly organized by Farrell and Odi, several faculty and program directors indicated that they did not see the Black Lives Matter conversation as their responsibility. Leah describes the differential responses of faculty and administrators to the issues:

> There were a lot of students who could not concentrate, they could not focus. Because they were black students who did not feel safe for their family members and they did not feel that Lehigh was a safe space. . . . There were a couple of professors and interim directors of different programs who kept arguing that . . . bringing up the Black Lives Matter conversation is not their responsibility because they teach chemistry or they teach bio. . . . And one student spoke up and said, "I am an engineering student, and if I cannot focus because thinking about the unarmed black men being murdered by police, you should take that into account. That is part of my education. You can't ask me to take this 'calc' exam where that is not my priority when my life is at risk." And there were professors, faculty and administrators who really did not see that as their place. And that was one of the major problems. What are the administrators doing to ensure that their faculty are on board with the whole diversity plan at Lehigh?

Building an Inclusive Campus Climate: A Work in Progress

Despite significant structural changes underway at Lehigh, members of nondominant groups may not yet perceive that these changes have had a measurable impact on the campus environment. The campus climate survey was conducted in the fall of 2015 and will be re-administered in 2016 and 2017 with the goal of fostering dialogue about diversity and inclusion and identifying strategies for improving the climate.[96] With a response rate of 41.8 percent, survey findings indicate that 48 percent of respondents found that Lehigh's intergroup climate is "positive" or "very positive," while women, minority students and employees, and lesbian/gay/bisexual/queer/ (LGBQO) were less likely to rate the climate as "positive" compared to their counterparts.[97] Improving the campus climate was rated as an extremely important goal by 33 percent of the respondents, with employees more likely to view this as an important goal than students.[98]

Ted Morgan, Professor of Political Science, indicates that the university needs to create a climate so that "all students generally really, truly feel welcome, like it's *their* university." Morgan, a white male, identifies two major factors affecting the campus culture and student life: the cohort of students traditionally attracted to Lehigh and the Greek system. Fraternities and sororities have often served to reinforce class-based, race-based, and male sexist forms of privilege within predominantly white college campuses.[99] Morgan summarizes the impact of these two factors on the campus culture:

> one [factor] is probably related to the cohort of students that Lehigh attracts as applicants—a certain number of them bring with them a kind of ignorance about diversity issues, about different races, different cultures, and probably bring some related attitudes. But the second piece that is very important is the Greek system that has been traditionally very dominant at Lehigh, that provides the sort of insulated environment within which these attitudes can be nurtured—an environment that poses an obstacle to the university's ability to intervene and integrate students into the larger university community.
>
> The Greek system really dominates the social life of students, the parties that the fraternities put on, often nowadays off

> campus, tend to focus early on first-year students and pull them into the system and a way of thinking about life at Lehigh and their futures. Much of which can be pretty detrimental. There are some pretty nasty activities and attitudes that occur within some fraternities, both in terms of gender or sexuality and in terms of race. But in their relative isolation and ascriptive practices, fraternities can create a climate that makes it difficult for some who are uncomfortable with these activities to speak up, thus perpetuating these practices. . . . And the Greek system has a very strong standing among alumni. A lot of alumni support the Greek system.

Morgan also notes that while many students may be appalled by overt racism, many are less politically and socially sensitive, not cognizant of cultural differences, and unaware of the workings of institutional racism. As he explains:

> I think a certain number of those students are sort of oblivious to cultural differences related to racial cultures. And probably not very attuned to the ways in which institutional racism works. Maybe a larger portion or very significant portion doesn't seem to be politically or socially sensitive at all. They just go about their lives. Some of them harbor attitudes or have peer support for attitudes or peer influences . . . that come out in sometimes overtly racist ways, and some other times in just kind of distance and coolness toward minority students.
>
> Given that minority students are only a very small proportion of the student population, it's easy to see why they feel like they're sort of a very marginalized subculture in the institution as a whole. In many ways it doesn't feel very welcoming to them. And having taught here when there were even smaller percentages of minority students, I have seen many times how this can be a disheartening experience.

Morgan's observations are shared by James Peterson who describes a generally collegial culture and notes the significant progress Lehigh has made in diversity and inclusion. Peterson also indicates that minority faculty sometimes find the culture less welcoming in comparison to their white counterparts:

> The university has made great progress over the time in which I have been at Lehigh. But it still has a *long* way to go to reach its own institutional goals around these issues.
>
> For the most part, [the culture] is not hostile, for the most part it is collegial. There are still pockets of resistance particularly around the influence of hiring faculty of color. And I am saying this from my own personal experience in hiring faculty of color. What I am also finding is that sometimes our faculty of color don't feel valued in the same ways that their white counterparts feel valued. I am even seeing some areas . . . where equity is going be a challenge for us going forward if we do not become more proactive about being inclusive and being aggressive about retaining the faculty that we hire.

The Student View of Campus Culture

From a student perspective, the internal dynamics among faculty, high-level administrators, and staff are noticeable and sometimes polarizing factors in the campus culture. Leah cites internal tensions and a relative devaluing of the input of Student Affairs as harmful to the university's ability to address the needs of diverse students:

> And there's a big strain with faculty and Student Affairs. A lot of the faculty . . . don't see it [diversity] holistically, see it as the job of Student Affairs people and they don't even value the Student Affairs staff. So there is that tension on campus between faculty and Student Affairs. During meetings, Student Affairs has the most information, because they are constantly spending time with students outside of class, they see students at night, and they . . . help students out and they know the students' needs. But when they communicate that with faculty they [the faculty] completely shut that down. . . . there is a gap between faculty, staff, and administration . . . I think staff is in the best position to move forward with diversity. Not the administrators. They are not there with the students on a day-to-day basis. They

sit in their offices; they go to meetings every hour. They are not best equipped to talk to students or to really move forward and create a plan that will work and they need to bring more Student Affairs staff to the table if they are willing to make Lehigh a better place.

Leah's critique identifies the disconnection of administrators and faculty with the lived experiences of diverse students on campus. While administrators are mired in back-to-back meetings and the faculty typically view student life as outside their purview, attention is often not paid to Student Affairs staff who work with students on a day-to-day basis. In her view, valuing Student Affairs and listening to students are essential steps in building a workable diversity and inclusion plan. Leah that she is awaiting more action on the part of administration to ensure that diversity plans trickle down to students:

> Whenever our presidents or provosts . . . put out these plans and these statements and these letters, I really take them with a grain of salt. Lehigh is a business school and it knows how to market itself to the world to attract students as customers. I just take it with a grain of salt. I just need to see more action ongoing or even on a month to month basis updating students, letting students know what's going on, how he is going about this plan. . . . another huge problem with diversity is that oftentimes people only look at the president or the provost or a couple of leading people, but it is the whole organization as an institution. So there's a lot of academics who trickledown the effects on students and staff, and they are never held responsible.

Leah's observations reflect the tendency for diversity plans at many institutions to become rhetorical statements that are not translated into concrete actions, processes, and the day-to-day experiences of students, staff, and faculty within an institution. The gap between espoused and enacted diversity statements can affect student success and lead to a sense of disillusionment.[100] As Monique, the minority department chair cited earlier, explains, staff may feel disempowered and unable to speak up without repercussions:

> The staff do not feel as engaged or feel that they have a voice. They feel that there would be repercussions if they spoke up [voiced their concerns] with any administration and the administration of student affairs particularly, so there's a bit of disengagement between the administrators, the staff and faculty.

Due to the hierarchical working conditions for staff and lower-level administrators, diversity strategic planning will benefit from empowering these stakeholders to have voice and contribute to the change process without fear of retribution.

Making the Academic Case for Diversity

In identifying the further diversification of faculty and staff as a major institutional objective, perceptions of a potential conflict between diversity and academic excellence still persist in formal processes such as hiring and promotion. This argument is familiar to hiring authorities in academic institutions across the nation.[101] When a narrow definition of excellence and academic quality permeates organizational processes, such perspectives can limit student educational opportunities and the process of educational transformation.[102] In counteracting the argument that diversity may involve lowering standards, Henry Odi emphasizes that diversity, inclusion, *and* excellence need to be part of the same language and the same sentence and explicitly linked to creativity and innovation:

> Some people do believe that diversity in their mind may be lowering some standards and my perspective is that diversity and inclusion and excellence are all related because if we don't apply the whole concept of Inclusive Excellence in the work that we do to advance and sustain diversity and inclusion, then it's not worth the effort. Diversity applies to everything we do, both in the classroom and outside the classroom. . . .
>
> So again, I think that the fabric of all of this is that diversity, inclusion, excellence must be used in the same language, in the

> same sentence. And also as when we think about excellence we think about creativity and innovation.

Odi explains the critical importance of making the "academic case" for diversity in terms of how inclusion generates value for the university. He describes how building an inclusive climate in the academic departments will enable faculty to do their best work due to the value added through diverse scholarly perspectives:

> you can have all diversity that you want in any organization, but if that diversity, or the activities around to explain diversity and folks who are part of that do not feel included, if inclusion is not part of that equation, then the diversity really will not be generating value. Because you tend to then have . . . an environment that operates in silos. [For example], people who have some research activities may not think of the value of interdisciplinary programs.

To counteract the pushback that can occur when opening the conversation with the term "diversity," Odi prefers to lead with the concept of inclusion:

> There is a sense that inclusion tends to open the door and when that door is open figuratively, then you have an opportunity to get into the value of diversity. Leading with diversity tends to raise some walls, so to speak, in some people's minds, and they tend to "diss" it and not want to be part of the conversation. But when you approach the conversation from the perspective of inclusion, it tends to be more inviting, and then you have more opportunity with the audience to engage in value of diversity. And through that conversation . . . you can engage in the fact that this also means inclusive excellence.

Nonetheless, Odi experienced substantive resistance when launching an initiative with the provost and deans designed to strengthen diversity in the university's faculty search processes:

> I have gotten a lot of resistance along those lines. Five years ago, the provost and I and the deans started an initiative

> around preparing search committees, having the search committees go through workshops addressing unconscious bias, where we advertise, why we advertise, . . . why search committees should be better prepared for . . . assessing and screening candidates, why should they be advertising in communities outside of their respective genres to make sure that the pool is diverse, all of this has attracted resistance. Faculty have their own mindset that is. . . . how they interpret diversity. . . . Are you expecting us to lower the standards now? They interpreted diversity to mean lowering standards. It's not attracting excellence.
>
> Diverse candidates do attract excellence. So we were insistent about pushing this forward and making the case that it is important for the university. This is important for every member of this community if we are going to attract the best and the brightest from diverse backgrounds, we have to reflect that.

These efforts initiated by the Provost's Office and the deans are designed to build a comprehensive vetting process that ensures the consideration of diverse candidates and builds greater accountability for diversity hiring.

A bright spot on the university's diversity horizon was the award in 2010 of a five-year National Science Foundation ADVANCE Institutional Transformation grant. The grant is designed to increase the advancement of women in academic science and engineering and was specifically focused on building on interdisciplinary strengths to enhance the recruitment and retention of women faculty in STEM fields.[103] It has enabled the university to strengthen training for search committees, hold workshops on implicit bias, and address a range of issues related to recruitment and hiring. Kristen Jellison, a white female faculty member in the College of Engineering, has served as Faculty ADVANCE Director for the past four years. Under her leadership, the program is now institutionalized at Lehigh and a center focused on female faculty funded by the university. Jellison describes the obstacles she faced in the early years of the ADVANCE grant when providing workshops and breakfast meetings and finding it difficult to attract participants:

> Initially there was a lot of resistance. We would have meetings to talk about best practices in recruitment and we would receive a lot of pushback from people questioning why this was important. People interpreted it as lowering our standards, just to fill quotas. The responses were really upsetting.

Yet as time went on, a transition began to place. As she explains:

> We still have work to do, but we are making progress. Now when we have these sessions, people don't really question why we are having them. I think people are starting to understand the value and why it's important. We may debate about what's the most effective way to do things to support diversity and be inclusive. But I rarely hear people question why we need to be inclusive anymore.

One of the successful best practices resulting from the ADVANCE grant was the creation of a male advocate program with 10 male advocates and 30 allies serving as advocates for female faculty in STEM. James Peterson offers an explanation regarding the barriers to diversifying the faculty in science, technology, engineering, and math (STEM) fields:

> One of the reasons why the School of Engineering and Business/Economics sometimes lag behind (and this is true nationally), is because of a sense that it is more difficult to diversify certain disciplines due to their traditions of being havens for white male scholarship. I think we have to challenge those mythologies and remember that if there are not enough people in the pipeline to help us meet our diversity goals, then I don't think that is the end of story. I think we have to also say, we are a university and we have the capacity to train women and people of color in those disciplines. We have to say, what are we doing to contribute to the pipeline through our own graduate programs?

Interdisciplinary Learning as a Prism for Diversity and Inclusion

As mentioned earlier, one of the differentiating strengths at Lehigh is the focus on interdisciplinary curricular content as an essential aspect of student learning. The university's interdisciplinary approach to undergraduate education and insistence on incorporating interdisciplinary perspectives in all major, minor, and certificate programs offers an innovative approach to attainment of the aims of a liberal education. Interdisciplinary studies emphasize integrative knowledge by drawing on the insights, methods, and content of multiple field to provide a horizon-expanding education that prepares students to address complex problems as leaders and responsible citizens.[104] James Peterson explains how interdisciplinary work contributes to diversity learning:

> And one of the reasons why interdisciplinary work is an important feature both of our investment in globalization and preparing students for the globalized world, *and* of understanding how D & I [diversity and inclusion] works is because the actual structure of interdisciplinary studies is inclusive. If you are interdisciplinary, you have to be thinking about multiple disciplines at the same time and essentially delve into how we think about diversity and inclusion. Whether that is religion studies which is interdisciplinary or Africana studies or Latin American studies or Asian studies, you are talking about academic disciplines that actually can serve as formal models for what diversity and inclusion can be, particularly in the academic space but hopefully also models in the university space more broadly construed.

A recent initiative in cluster hiring awarded the first such interdisciplinary hires to Africana Studies, resulting in the appointment of five new faculty members. Approved by the Board of Trustees as well as a universitywide committee, the initial cluster hiring initiative signaled positive changes for diversity and inclusion in the educational process. As Peterson observes, "I think every level of the cluster hiring process showed an institutional commitment to diversity through the Africana Studies program."

Ted Morgan describes the Africana Studies program as "a very vital, very visible, and maybe programmatically the most visible thing happening on campus":

> Not only did it start to change the racial makeup of the faculty, but it has been an important catalyst for Lehigh to begin to change. I think the university is becoming a lot more conscious of how they have been negligent for a long time and how they really need to do some things differently. I think at the same time the university historically has been *very* loathe to alienate the alumni, and I think the alumni historically have been a sort of conservative, corporate and Greek-system-supporting population, that has made change on things like this move at a glacial pace generally.

As Morgan's comments suggest, interdisciplinary programs are a significant lever triggering cultural change. Nonetheless, a tendency may exist for some white faculty members to devalue the scholarship of minority faculty members whose work contributes to interdisciplinary, multicultural, feminist, and minority studies programs.[105] Fred Bonner points out that academe can offer "only tepid affirmation and support" for minority faculty engaged in scholarship related to diversity.[106]

Dimensions of Diversity Organizational Learning at Lehigh

Coupled with the impact of interdisciplinary learning, one of the significant strengths in diversity education at Lehigh is that programs are generated internally and draw on the expertise and scholarship of faculty. James Peterson describes with enthusiasm the role of Lehigh's own faculty in advancing diversity education:

> I think what is most exciting to me is the things we are doing internally. We have been doing teacher trainings on inclusivity

> that are run and facilitated by Lehigh faculty. I think that's the most compelling way to advance along the trajectory of education around D & I [diversity and inclusion]. But I also think that we have done some great programming, again internally . . . on our own, the hiring of Africana faculty, all these things together. . . . help to move the needle. And . . . once you start to make progress, if you are smart, you will realize how much progress you still have to make. I think that's why we continue to make some progress. I am happy to be a campus leader around some of these issues. But I also see that we have a lot of work to do.

Other facets of diversity organizational learning programs at Lehigh include the effort to tackle difficult questions through campuswide conversations, events, and symposia on key issues. For example, following the divisive 2016 presidential election, several hundred members of the Lehigh community gathered to conduct a peaceful rally that emphasized inclusion both on campus and in the nation.[107] A series organized by the Education and Dialogue group of the Council for Equity and Community titled "Tackling Tough Topics Together" examines difficult issues such as thinking about the identities that define individuals and the challenges posed by these identities.[108] In past years, the Provost's office has worked with several external groups to deliver diversity-related seminars and interactive theater programs. While these offerings have focused more on faculty and staff, the emphasis has now shifted to the development of in-house education such as working with Lehigh's Theater Department to build scenarios based on issues identified by specific groups in the campus climate survey.

Looking Forward

As Lehigh moves forward in its diversity planning and education programs, the university has made substantive progress over the last ten years in faculty and student diversity. With an 18 percent increase in the number of faculty, the number of full-time African American faculty doubled from 9 to 18 (100 percent increase), Latino/a faculty increased from 8 to 18 (125 percent increase), and Asian American faculty increased from 44 to 68 (55 percent increase). On the student side, white student enrollment has dropped 15 percent while Latino/a enrollment has increased 153 percent, Asian American enrollment has increased 58 percent, and African American enrollment has increased 23 percent.

Characteristics of the current Lehigh environment that offer significant promise for further advances in diversity and inclusion include 1) academically focused presidential leadership with an emphasis on cultural diversity and global student learning; 2) strategic planning and goal setting with attention to the need for demographic diversity of faculty, staff, and students; 3) structural changes that expand the diversity infrastructure across the university; and 4) a concerted array of intellectually rich programs focused on diversity and inclusion that emanate internally from the engagement of faculty, staff, administrators, and students.

While university leadership has taken concrete and measurable steps to strengthen the diversity infrastructure through strategic planning, curricular innovation and other organizational learning initiatives, there is still work to be done to build a more inclusive learning environment and to create a holistic ecosystem for diversity. Summing it up, Henry Odi describes the cultural change and diversity learning process at Lehigh as evolutionary:

> I use the metaphor of a journey. It is a journey, it is not an end game that we can say we have arrived: there is no such thing, because the community and country continue to evolve and change.

Case Study V
Moving from Compositional Diversity to Inclusive Learning Practices

The University of Maryland at College Park

This case study examines progress in diversity organizational learning at the University of Maryland at College Park (UMD), a large, highly diverse public doctoral research university that is the flagship campus of the University of Maryland System (USM). UMD is located just eight miles from Washington, D.C.

The University of Maryland at College Park's (UMD) fall 2015 undergraduate enrollment included students who are 12.8 percent African American, 16.2 U.S. Asian American, 9.3 percent Hispanic students, 4.1 percent of two or more races, as well as an international student contingent of 4.1 percent.[109] Minority students make up 42.6 percent of the undergraduate population and an even larger share of graduate students, with 31.2 percent of these students coming from foreign countries.[110] A slight increase has been seen in the hiring of diverse tenure-track faculty from the United States, moving from approximately one-fifth of the faculty in 2008 to 23 percent by 2014.[111] Nonetheless, the percentage of African American faculty actually declined from 5.1 percent to 4.4 percent, while the representation of U.S. Asian/Pacific Islander faculty increased from 11 percent to 14.1 percent.[112] Foreign faculty comprised 4.3 percent of this group in 2008 and 3.6 percent in 2014.

Historical Legacy

The state of Maryland has a long history of racial segregation. The Civil Rights project at the University of California at Los Angeles (UCLA) recently found that Maryland had moved to the top of the list among states where African American and Latino K-12 students are most segregated due to residential resegregation in the suburbs.[113] UMD first admitted an African American male student in 1951 and a female in 1955. President Harry Clifton "Curly" Byrd fought to ensure that the College Park campus was segregated by making "the branch campuses adequate and maybe the Negroes could be content."[114] Only a small percentage of African American students were enrolled until the late 1970s with an undergraduate African American enrollment of 7.4 percent in 1978.[115]

In the late 1970s, UMD established the Benjamin Banneker scholarship under an Office of Civil Rights order of the U.S. Department of Education. According to William E. Kirwan, President of the University of Maryland from 1998 to 2002 and later Chancellor of the University System, the scholarship sought to dismantle the physical and psychological effects of the state's segregated system of higher education.[116] The scholarship provided full four-year scholarships to 30 or 40 top African American students. Daniel J. Podberesky, a Hispanic student, sued the university claiming he had been denied the scholarship since he was not African American.[117] In 1995, the Supreme Court denied further review of the Fourth Circuit federal appeals court ruling finding the race-based scholarship in violation of the Fourteenth Amendment's Equal Protection clause.[118] In foreclosing public-funded, race-based scholarship, this decision had ripple effects throughout the Fourth Circuit in the states of North Carolina, South Carolina, Virginia, and West Virginia.[119]

The Path to Diversity Progress

As can be seen from the furor over the Banneker scholarship, the road toward diversity progress at UMD has not always been a smooth one. In 2009, several hundred students marched in protest of the firing of Dr. Cordell Black, an African American associate provost for equity and diversity who had held the position for 18 years.[120] The provost, Dr. Nariman Favardin, stated the change was for budgetary reasons and the position would become part-time. Nonetheless, the salary saving appeared to only be about

$10,000 since Dr. Black returned to a tenured faculty line.[121] Favardin later assumed a post as President of Stevens University.

In 2010, Dr. Wallace D. Loh took office as president of the University of Maryland, one of the small number of Asian or Asian American presidents heading American institutions of higher education. According to a study by the American Council of Education, in 2011, 13 percent of college presidents were minorities and only 2 percent Asian American.[122] Born in China, Loh moved with his family to Peru, and later emigrated to the United States where he became a naturalized citizen.[123]

In 2010 UMD published a ten-year diversity strategic plan titled "Transforming Maryland: Expectations for Excellence in Diversity and Inclusion" signed by President Loh that had been formulated collaboratively by the Diversity Plan Steering Committee in consultation with university stakeholders and committees over an 18-month period. The plan focuses on impact, leadership, and excellence and explicitly identifies diversity as a "cornerstone" of excellence.[124] Three tenets underlie the plan's relation to Inclusive Excellence: 1) that diversity contributes to the excellence and vitality of the educational process, 2) that students need exposure to different perspectives to work successfully in a diverse workplace and global society, and 3) as a state institution, all citizens need access to the "transformative experience" provided through outstanding higher education programs.[125] The plan acknowledges UMD's "deplorable practices of discrimination" in terms of its founding in 1856 by 24 trustees, 16 of whom were slave owners.

UMD's diversity strategic plan provides concrete goals and strategies in six core areas: leadership, climate, recruitment and retention, education, research and scholarship, and community engagement. At the time of the plan's development, the university had a special assistant to the president for diversity and an associate provost for diversity, but no cabinet-level diversity position.

Although relatively late to address the need for structural diversity leadership in comparison to peer institutions, one of the plan's key recommendations was the hiring of a chief diversity officer to report to the president and sit on cabinet, reformulation of a campuswide Diversity Advisory Council to replace the Equity Council, and creation of an Office of University Diversity. Dr. Kumea Shorter-Gooden, an African American female, was subsequently hired as the university's inaugural Chief Diversity Officer and Associate Vice President, reporting to the Provost with a dotted line to the President. Dr. Shorter, a licensed psychologist, holds a Ph.D. in Clinical/Community Psychology from the University of Maryland at College Park. Her position does not have tenure and supervises six staff. The staffing design reflects a strong emphasis on education and training with both a director and manager of education and training programs and two education and training specialists.

In November 2015 students at Maryland joined the wave of protests sparked by the incidents at the University of Missouri at Columbia, calling for the renaming of Byrd Stadium, the football stadium named for Harry Clifton "Curley" Byrd, the avid segregationist who had denied African American students admission to the campus.[126] Subsequently, President Wallace Loh recommended a name change to the University System of Maryland Board of Regents, writing that the symbolism portrayed in the "front porch" of the institution "conveys a racial message in plain sight."[127] With the approval of the Board of Regents, the stadium was renamed the Maryland Stadium.

A highly publicized incident roiled the campus in 2015 when an email from an undergraduate member of the Kappa Sigma fraternity written in 2014 was posted on the internet in March 2015. The email included racial epithets and graphic remarks with expletives indicating the need to forget about consent in having sex with women during rush week, but not including women of certain races.[128] About 100 students protested outside the fraternity following a sit-in at the student union and President Wallace Loh joined the sit-in to express his solidarity.[129] On April 1, 2015,

President Loh announced the results of an investigation by the campus and Prince George's County Police and the university's Office of Civil Rights and Sexual Misconduct, stating that while the email was "profoundly hurtful" as well as "hateful and reprehensible," it did not violate university policy and was protected by the First Amendment's guarantee of free speech.[130] Following the response, a meeting with the president was requested by students, faculty, staff, and parents.[131] Throughout the controversy, President Loh tweeted his concern over the dividing line between free speech and hate speech.[132] The decision not to expel the student was viewed as part of a rehabilitative and educational effort, with the student apologizing and performing community service as well as individualized training in diversity and cultural competence.[133]

The issue of free speech and academic freedom versus hate speech has torn a number of campuses apart. President Wallace Loh made a decision in line with the investigative findings that was viewed by weak by some critics and contrasted with his own personal views. This debate continues on campuses around the nation on issues regarding academic freedom, free speech, and racially charged or discriminatory remarks by students and faculty members. Kimberly Griffin, Associate Professor of Higher Education who studies campus climate, notes that one positive outcome of the debate was the opening of a larger conversation about the institution's response to acts of marginalization and oppression:

> The good thing that came from it was that it started a larger conversation around how institutions can and should be responding to those things. In some ways we are trying to create inclusive campus communities, [but] that doesn't mean that we can prevent every act of oppression or marginalization that takes place on campus. But there is a lot to be said about how campuses respond to those things and how we come together as community to suggest that those things aren't right, and that they exclude important people and they marginalize them and that they are racist and sexist, and all those other challenging things.

As in other large public universities, the creation of systemic, structural processes that foster inclusion is still a work in progress. This case study illustrates the specific challenges faced by institutions that have laid substantial groundwork for diversity and inclusion in the mobilization and implementation phases of cultural change, but still are working to institutionalize systemic diversity learning initiatives across the broad, decentralized terrain of a large, doctoral research university.

Inclusive Excellence and the Ongoing Process of Diversity Learning

While UMD has made significant strides in its efforts to diversify the administration and the faculty, the opportunity for students to find diverse role models in their fields of study still may be limited. As Johanna de Guzman, a Filipino American undergraduate and former Director of Diversity for the Student Government Association explains:

> It was just a year ago, my junior year, that I had a professor that looked like me, that I could see myself in, where I could see my ethnic background in and it was because I was taking an Asian American studies course. A lot of students of color do say that it is hard finding administrators and professors that they can relate to on a background basis, where it's like you came from somewhere like I did understand or you understand my ethnic background or religious background.

And even with the increased attention given by the university to building compositional diversity, the more complex work of inclusion is still a work in progress. This work, as Shorter-Gooden explains, involves the creation of an environment in which diversity is embraced and diverse students, faculty, and administrators can affirm their own identities and be welcomed for who they are:

> And there is a lot of attention to how to build on the compositional diversity but maybe even more significantly, how to create

> more inclusion. We know that there is a difference between having people inside the campus gates and having spaces where people feel fully embraced, not just tolerated, where they feel well utilized and engaged, where they feel like they can bring their whole self to school and to work. I think we do well at times, and not as well at other times. So inclusion, of course is more complicated than compositional diversity and we know that we have lots of work to do.

In articulating the ways in which the Office of Diversity and Inclusion approaches diversity organizational learning, Kumea Shorter-Gooden emphasizes the critical need to name realities while not simultaneously creating the notion that there are "good guys" and "bad guys" or creating a sense of individual or collective guilt for historical legacies of exclusion:

> We always try to work to create spaces where people who are vulnerable can be heard, and bring their whole selves to the campus. Where we also name realities: racism IS alive and well; heterosexism does continue to exist, sexism as well; trans people do experience trans phobia. It's really important that we not act like these things are just a matter of opinion. But there is documented evidence that there are systematic disparities based on race, religion, gender, and so forth. It's important to name these things while making sure that we are not conveying that there's a bad guy. So one of the things that I think that we do pretty well, the whole tenor of the work that my campus leads, and I can say this is true across campus, is not one that says there is a set of good guys and there is a set of bad guys, and by the way the bad guys are white, male, straight, Christian, Cis.

Furthermore, she warns against the risk of diversity programming as the ultimate solution or an end in itself, without considering necessary structural or systemic change. In such cases, events could devolve into politically correct cheerleading, without addressing underlying realities. As she explains:

> There is a risk that diversity events seminars will become political correctness fests and that people will cheerlead and feel affirmed which are not bad things but where not much openness or learning happens. Another risk is that, painting it most cynically, that we have got mostly students a few staff and faculty busy talking and doing things and having important engaged discussions about diversity and feeling like that's making a difference, but meanwhile nothing structurally in the university is changing. It's a risk of having diversity events or training as the only or ultimate solution. In contrast to having it as part of what one does that help people think about some of the structural or systemic change that needs to occur.

Nodes of Success in Campus Diversity Education

Over the past five years, UMD has moved well beyond the mobilization phase of diversity culture change and built a substantive infrastructure with diversity officers in each of the schools, colleges, and divisions and a campuswide Diversity Council. Yet as with many large institutions, the university is highly decentralized with substantial power residing within the colleges. Kerry Ann O'Meara, Professor of Higher Education, underscores the nature of organizational learning as a constant process, noting the need for greater exposure among some sectors of the university to contemporary forms of discrimination and the resulting impact on opportunity structures for diverse individuals:

> I think a wider spread understanding of and appreciation for implicit bias and micro-aggressions, and the experiences of people who face the isms—like sexism, racism, homophobia, discrimination, discrimination by ability, or by international kinds of issues [is needed]. I think just an appreciation for the experiences of others and what that means in terms of opportunity structures. I think there are just different levels of knowledge about those things and it's beginning to spread. But we've got more work to do. There are some sectors of the university have not been quite as exposed to this and where there is still

> inequality that happens and missed opportunities to help people succeed and probably understand about why some folks are not doing well and others are.

While the campus is highly diverse from a structural perspective, student comments still suggest some degree of student self-segregation among students of similar ethnic, racial, religious, or cultural backgrounds.[134] A barrier to creating cohesive diversity programming across the multitude of affinity-based student groups is the lack of physical space and the tendency of identity groups to congregate with similar others. Johanna de Guzman describes the challenges of physical space that can preclude alliances and collaborative work on diversity among student groups:

> Walking on the campus, you are greeted with a bunch of different cultures, a bunch of different ethnicities, religions, just a bunch of people who you can tell are from just different backgrounds. However, I don't think the campus is built to hold this many diverse groups on campus, so there is not enough space for cultural groups or religious groups or LGTQUIA groups. And we tend to be separated on campus only because of where the offices are. So there is no central place for all of us to meet, for all of us to mingle, to work together, and to have conversations on events that we could plan or initiatives that we could push for. I think the campus is moving in the right direction. But I also do think that we have a long way to go.

De Guzman identifies the role of the Multicultural Center as a node of campus success in promoting the inclusion of and advocacy for minoritized students:

> the creation of that space for students and the people who work there: they advocate for their students every single day and students feel as though I have a place where administration is on their side and it doesn't fall into the politics of "we can't do this." It's "I can do this" because it's my job to advocate for you.

Kimberly Griffin offers a research-based perspective on the ways in which identity groups can provide emotional and psychological support on a predominantly white, heterosexual campus:

> There are times when students tend to congregate when people who are like themselves; I think we don't think about the times when white students congregate with others that look like themselves. We tend to focus mostly on students of color. . . . Research suggests that while we may see students spending time who look like themselves, students with marginalized backgrounds, students of color, LGBT students . . . a lot of times that's because they feel marginalized in larger spaces that they very rarely see other people who look like them, that they are experiencing forces of oppression and micro-aggressions that make them feel like they don't fit and they don't belong, so that they need the support of folks who look like them as they are engaging with folks from different backgrounds or engaging across difference.

For this reason, Griffin notes research showing that minoritized students spend more time engaging across difference than white students and emphasizes the "both and" rather than "either or" of such engagement.

Building Synergy for Diversity Learning Across Institutional Silos

In the view of Kumea Shorter-Gooden, UMD is at an advanced implementation stage in terms of finding ways to bridge institutional silos and create systemic programs across administrative, faculty, staff, and student divides. While she underscores the positive support of university leadership for such efforts, she notes that given the conservative nature of educational institutions, the national climate has generated further impetus for change:

> I think institutions are not good at transforming themselves except under duress. I would say that's the hard part for us and I think for others. Most institutions are conservative, and they do

> what they have been doing, and in some ways, I think that this particular time in the nation provides that duress that will help us to move further.

The initiatives underway at Maryland build on the existing diversity infrastructure and include gap assessment, curricular learning, and universitywide events that promote diversity organizational learning.

Gap Assessment

To provide a basis for further planning and gap assessment, a comprehensive Gallup survey of faculty and staff was conducted regarding levels of engagement and experiences of diversity. Debriefing sessions were held to engage staff and faculty around issues of diversity and inclusion and to assist the most senior levels of leadership in collectively planning. Given the existing infrastructure of diversity officers in each college or division, the survey will help the establishment of more short-term, campuswide goals in conjunction with the long-term goals of the diversity strategic plan.

Curricular Change

One of the most prominent successes in disseminating diversity learning to undergraduate students has been the establishment of a two-course diversity requirement as part of the newly revised General Education program. This requirement involves the selection by students of either two Understanding Plural Societies (UP) course or one Cultural Competence course. Built through faculty collaboration and use of focus groups, a rubric for cultural competence identifies specific learning outcomes designed to foster the awareness, knowledge, and skills to communicate across cultural differences both in and outside of the classroom.[135]

Coupled with this initiative, the Office of Diversity and Inclusion and the College of Education have collaborated to provide credit-bearing Intergroup Dialogue courses to students. In 2013–2014, a

Cultural Competence Course Development Project (CCCDP) was launched by the Office of Diversity and Inclusion in partnership with undergraduate studies to work with 21 faculty members in reworking existing courses to meet the criteria for Cultural Competence.

Diversity Organizational Learning

Like other CDOs, Dr. Shorter-Gooden describes the national context as escalating the number of requests for diversity training over the past couple of years. Events in Ferguson, Orlando, and Baltimore have brought questions of race to the fore and have affected members of the millennial generation, some of whom had been less aware of the interplay of power and privilege.

A prominent universitywide initiative designed to bridge organizational silos is "Rise Above -isms", a week-long event that seeks to engage the entire campus community in rising above stereotypes associated with racism, sexism, ableism, and other isms. The event offers micro-grants for programming that helps participants examine their own identities, assumptions, and biases in a safe and structured environment.[136] According to Dr. Shorter-Gooden, the event has been "very deliberate" in focusing on intersectionality and multiple identities and created safe spaces where people can be themselves, avoiding the tendency to subdivide groups based on single facets of identity. While successful in terms of attendance levels, she indicates that the program has had greater impact on students and some staff and faculty, with less impact on senior leadership.

Two other dialogue-based programs are directed toward strengthening communication across differences. The newly launched Maryland Dialogues on Diversity are focused on changing culture and creating a cultural norm of communicating with and learning from those who are different.[137] The first year's focus was on race and racism and intersectionality. The Words of Engagement program is an optional course that allows students to discuss historic areas of historical conflict such as issues of race and social biases.[138] Key attributes of these programs include their informal nature and emphasis on naming reality, working through conflict, and fostering intergroup communication.

Another major effort launched by the Office of Diversity and Inclusion with the support of the Provost was the creation of a pilot program to train search committees on inclusive search practices and the impact of implicit bias. Sixteen of the university's colleges are currently participating in the program.

Challenges on the Road Ahead

Given the existence of an infrastructure for diversity that addresses the decentralized landscape of colleges and schools, UMD has taken intentional steps toward building a more inclusive culture and institutionalizing diversity organizational learning. Like other institutions of higher education, Maryland has responded to student protests and addressed persistent symbols of the racial past such as through the renaming of Byrd Stadium. These protests continued in 2017 with the presentation of 64 demands by ProtectUMD, a coalition of 25 student groups, calling for greater transparency by the administration and expressing concern regarding the resignation of Kumea Shorter-Gooden as Chief Diversity Officer in January 2017. Dr. Roger Norrington, formerly a professor in the College of Education, was named as the new Chief Diversity Officer and Associate Provost in July 2017.

One of the principal challenges on the road ahead is to remain proactive in addressing structural inequality, responding to a series of racially motivated incidents on campus, and examining the ways in which social forces have impacted systems of power and privilege as replicated in the campus environment. Under Kumea Shorter-Gooden's leadership, the Office for Diversity and Inclusion led and took part in campus and community conversations on difficult social issues such as the Black Lives Matter movement, racial profiling, stereotyping, modern racism, and everyday forms of discrimination. Dr. Shorter-Gooden emphasized the importance of educating students on contemporary social issues and engaging with the community, stating:[139]

> If we try to educate our students more on matters such as this, we could be able to see our community [and the] university become more engaged with each other and the issues that matter to all of us.

While to date students have been a driver in the change process, active institutional attention and intervention will be needed to ensure sustained progress. Kimberly Griffin perceptively highlights the importance of being proactive in creating a sense of belonging and inclusion on campus, even in the absence of student pressures:

> I think it's easy to be really frustrated and really upset when students are protesting and they're upset about what's happening on their campus. Institutions often kind of react like deer in headlights: they're scared when students are protesting and that doesn't happen on our campus. And we tend to be really reactive rather than proactive.
>
> So I'm hoping that the next phases of organizational learning are really us being proactive, us thinking about what it takes to have an inclusive community; what different communities say they are missing and they're lacking and what they need; and what makes feel them like they are welcome and like they have a sense of belonging on campus; and how do we provide those things on the front end, and how do we cultivate environments that really support that without students having to protest.

Notes

1. Wilson, E. O. (1998). *Consilience: The unity of knowledge*. New York: Vintage Books.
2. Bezrukova, K., Jehn, K. A., and Spell, C. S. (2012). Reviewing diversity training: Where we have been and where we should go. *Academy of Management Learning and Education*, 11(2), 207–227.
3. Ibid.
4. Paluck, E. L., and Green, D. P. (2009). Prejudice reduction: What works? A review and assessment of research and practice. *Annual Review of Psychology*, 60, 339–367.
5. Kidder, D., Lankau, M. J., Chrobot-Mason, D., Mollica, K. A., and Friedman, R. A. (2004). Backlash toward diversity initiatives: Examining the impact of diversity program justification, personal, and group outcomes. *International Journal of Conflict Management*, 15(1), 77–102.

6. Ragins, B. R. (1997). Diversified mentoring relationships in organizations: A power perspective. *The Academy of Management Review*, 22(2), 482–521. Chun, E., and Evans, A. (2012). *Diverse administrators in peril.*
7. King, E. B., Gulick, L. M. V., and Avery, D. R. (2010). The divide between diversity training and diversity education: Integrating best practices. *Journal of Management Education*, 34(6), 891–906.
8. Ibid.
9. Ibid.
10. Kraiger, K., Ford, J. K., and Salas, E. (1993). Application of cognitive, skill-based, and affective theories of learning outcomes to new methods of training evaluation. *Journal of Applied Psychology*, 78(2), 311–328.
11. Roberson, L., Kulik, C. T., and Pepper, M. B. (2009). Individual and environmental factors influencing the use of transfer strategies after diversity training. *Group and Organization Management*, 24, 67–89.
12. Fisher-Yoshida, B., Geller, K. D., and Schapiro, S. A. (2009). Introduction: New dimensions in transformative education. In B. Fisher-Yoshida, K. D. Geller, and S. A. Schapiro (Eds.), *Innovations in transformative learning: Space, culture, and the arts* (pp. 1–22). New York: Peter Lang Publishing.
13. Ibid. See also Mezirow, J. (1991). *Transformative dimensions of adult learning*. San Francisco: Jossey-Bass.
14. Mezirow, J. (2009). Transformative learning theory. In J. Mezirow and E. W. Taylor (Eds.), *Transformative learning in practice: Insights from community, workplace, and higher education* (pp. 18–32). San Francisco: John Wiley & Sons.
15. Ibid.
16. See Illeris, K. (2014). *Transformative learning and identity*. New York: Routledge.
17. Green Fareed, C. (2009). Culture matters: Developing culturally responsive transformative learning experiences in communities of color. In B. Fisher-Yoshida, K. D. Geller, and S. A. Schapiro (Eds.), *Innovations in transformative learning: Space, culture, and the arts* (pp. 117–132). New York: Peter Lang Publishing.
18. Ibid.
19. For fuller discussion, see Taylor, E. W., and Jarecke, J. (2009). Looking forward by looking back: Reflections on the practice of transformative learning. In J. Mezirow and E. W. Taylor (Eds.), *Transformative learning in practice: Insights from community, workplace, and higher education* (pp. 275–290). San Francisco: John Wiley & Sons.
20. Rynes, S., and Rosen, B. (1995). A field survey of factors affecting the adoption and perceived success of diversity training. *Personnel Psychology*, 48(2), 247–270.
21. Beer, M., Finnstrom, M., and Schrader, D. (2016, October). Why leadership training fails-and what to do about it. *Harvard Business Review*. Retrieved March 7, 2017, from https://hbr.org/2016/10/why-leadership-training-fails-and-what-to-do-about-it
22. Ibid. See also Roberson, L., Kulik, C. T., and Tan, R. Y. (2013). Effective diversity training. *The Oxford Handbook of Diversity and Work*. Retrieved March 7, 2017, from www.oxfordhandbooks.com/view/10.1093/oxfordhb/9780199736355.001.0001/oxfordhb-9780199736355-e-19
23. Ibid.
24. Ibid.
25. Evans, A., and Chun, E. (2012). *Creating a tipping point: Strategic human resources in higher education* (ASHE Higher Education Report, Vol. 38, No. 1). San Francisco: Jossey-Bass.
26. See for example, Ulrich, D., Brockbank, W., Johnson, D., Sandholtz, K., and Younger, J. (2008). *HR competencies: Mastery at the intersection of people and business*. Alexandria, VA:

Society for Human Resource Management. See also Ulrich, D., Allen, J., Brockbank, W., Younger, J., and Nyman, M. (2009). *HR Transformation: Building human resources from the outside in.* New York: McGraw-Hill.

27. Ulrich, Allen, Brockbank, Younger, and Nyman. (2009). *HR Transformation.* See also Evans and Chun. (2012). *Creating a tipping point.*
28. Marchesani, L. S., and Jackson, B. W. (2005). Transforming higher education institutions using multicultural organizational development: A case study of a large northeastern university. In M. L. Ouellett (Ed.), *Teaching inclusively: Resources for course, department & institutional change in higher education* (pp. 241–257). Stillwater, OK: New Forums Press.
29. See for example, King, Gulick, and Avery. (2010). The divide between diversity training and diversity education. See also

    Roberson, L., Kulik, C. T., and Pepper, M. B. (2009). Individual and environmental factors influencing the use of transfer strategies after diversity training. *Group and Organization Management*, 24, 67–89.
30. Roberson, Kulik, and Pepper. (2009). Individual and environmental factors influencing the use of transfer strategies after diversity training.
31. Ibid.
32. Paluck, E. L. (2006). Diversity training and intergroup contact: A call to action research. *Journal of Social Issues*, 62(3), 577–595.
33. Bendick, M. Jr., Egan, M. L., and Lofhjelm, S. M. (2001). Workforce diversity training: From anti-discrimination compliance to organizational development. *Human Resource Planning*, 24(2), 10–25.
34. Samuels, D. R. (2014). *The culturally inclusive educator: Preparing for a multicultural world.* New York: Teachers College Press.
35. Kidder, D., Lankau, M. J., Chrobot-Mason, D., Mollica, K. A., and Friedman, R. A. (2004). Backlash toward diversity initiatives: Examining the impact of diversity program justification, personal, and group outcomes. *International Journal of Conflict Management*, 15(1), 77–102.
36. Dobbin, F., and Kalev, A. (2016). Why diversity programs fail. *Harvard Business Review.* Retrieved March 15, 2017, from https://hbr.org/2016/07/why-diversity-programs-fail
37. Paluck, E. L., and Green, D. P. (2009). Prejudice reduction: What works? A review and assessment of research and practice. *Annual Review of Psychology*, 60, 339–367.
38. *Leadership & Race: How to develop and support leadership that contributes to racial justice.* (2010). Retrieved September 24, 2016, from http://leadershiplearning.org/system/files/Leadership%20and%20Race%20FINAL_Electronic_072010.pdf
39. Stanley, C. (2017). Reflections on changing a university's diversity culture. *Insight Into Diversity.* Retrieved March 14, 2017, from www.insightintodiversity.com/reflections-on-changing-a-universitys-diversity-culture/
40. Lahiri, I. (2008). *Creating a competency model for diversity and inclusion practitioners.* Retrieved March 15, 2017, from www.conference-board.org/pdf_free/R-1420-08-RR.pdf
41. Worthington, R. L., Stanley, C. A., and Lewis, W. T., Sr. (2014). *National association of diversity officers in higher education: Standards of professional practice for chief diversity officers.* Retrieved March 15, 2017, from www.nadohe.org/standards-of-professional-practice-for-chief-diversity-officers
42. Ibid.
43. Chun, E. (2009). Dramatizing diversity. *Affirmative Action Register.* Retrieved March 15, 2017, from http://ednachun.com/Dramatizing%20Diversity%20-%20Chun.PDF
44. Cockell, J., and McArthur-Blair, J. (2012). *Appreciative inquiry in higher education: A transformative force.* San Francisco: John Wiley & Sons.

45. Williams, D. A. (2013). *Strategic diversity leadership: Activating change and transformation in higher education*. Sterling, VA: Stylus.
46. Ibid.
47. Cook, C. E., and Sorcinelli, M. D. (2005). Building multiculturalism into teaching development programs. In M. L. Ouellett (Ed.), *Teaching inclusively: Resources for course, department & institutional change in higher education* (pp. 74–83). Stillwater, OK: New Forums Press.
48. Beach, A. L., Sorcinelli, M. D., Austin, A. E., and Rivard, J. K. (2016). *Faculty development in the age of evidence: Current practices, future imperatives*. Sterling: VA: Stylus Publishing.
49. Ibid.
50. Ibid.
51. Schroeder, C. M., and Associates. (2011). *Coming in from the margins: Faculty development's emerging organizational development role in institutional change*. Sterling, VA: Stylus Publishing.
52. *Fellowship*. (2017). Harvard University. Retrieved March 16, 2017, from http://diversity.harvard.edu/pages/fellowship
53. For a listing of the principal components of academic leadership programs, see Gmelch, W. H., and Buller, J. L. (2015). *Building academic leadership capacity: A guide to best practices*. San Francisco: Jossey-Bass.
54. Gmelch and Buller. (2015). *Building academic leadership capacity*.
55. *UC Irvine diversity development program*. (2015). Retrieved March 20, 2017, from www.oeod.uci.edu/ddp.html

    Diversity & inclusion certificate. (n.d.). Florida State University. Retrieved March 20, 2017, from http://thecenter.fsu.edu/learning/diversity-inclusion-certificate. See also *Office of diversity & inclusion*. (n.d.). University of Pittsburgh. Retrieved March 20, 2017, from www.diversity.pitt.edu/education-training/diversity-and-inclusion-certificate-program
56. *Annual enhancing cultural competency conference*. (n.d.). The University of Kansas. Retrieved March 17, 2017, from http://diversity.ku.edu/annual-enhancing-cultural-competency-conference
57. *Diversity, equity & inclusion*: *Strategic plan*. (2016). University of Michigan. Retrieved February 22, 2017, from https://diversity.umich.edu/wp-content/uploads/2016/10/strategic-plan.pdf
58. Sellers, R. (2017, March 23). Personal communication.
59. Saul, S. (2016, September 6). Campuses cautiously train freshmen against subtle insults. *The New York Times*. Retrieved September 20, 2016, from www.nytimes.com/2016/09/07/us/campuses-cautiously-train-freshmen-against-subtle-insults.html?_r=1
60. O'Connell, S. (2016). *Clark president defends university's diversity training*. Retrieved September 24, 2016, from www.telegram.com/news/20160909/clark-president-defends-universitys-diversity-training
61. Ibid.
62. New, J. (2016, August 30). Renewed diversity push. *Inside Higher Ed*. Retrieved March 17, 2017, from www.insidehighered.com/news/2016/08/30/protests-racist-incidents-lead-more-multicultural-programs-campuses
63. Brown, S. (2016, May 15). Required for all new students: Dialogue 101. *The Chronicle of Higher Education*. Retrieved March 17, 2017, from www.chronicle.com/article/required-for-all-new-students-/236440
64. Hurtado, S., and Halualani, R. (2014). Diversity assessment, accountability, and action: Going beyond the numbers. *Diversity and Democracy*, 17(4), Retrieved February 14, 2017, from www.aacu.org/diversitydemocracy/2014/fall/hurtado-halualani

Halualani, R. T., Haiker, H. L., Lancaster, C., and Morrison, J. H. (2015). *Diversity mapping data portrait.* Retrieved July 5, 2015, from https://csumb.edu/sites/default/files/images/st-block-95-1429229970817-raw-csumbdiversitymappingdataportrait.pdf

65. Halualani, R. T., Haiker, H., and Lancaster, C. (2010). Mapping diversity efforts as inquiry. *Journal of Higher Education Policy and Management*, 32(2), 127–136.
66. Ibid.
67. Halualani, Haiker, Lancaster, and Morrison. (2015). *Diversity mapping data portrait.*
68. Halualani, R. (2013). *Diversity and inclusion at Penn State: Where are we now and what's next?* Retrieved March 13, 2017, from www.opia.psu.edu/print/3233
69. Ibid.
70. Samuels. (2014). *The culturally inclusive educator.*
71. *Lehigh University.* (2016). U.S. News & World Report. Retrieved December 27, 2016, from http://colleges.usnews.rankingsandreviews.com/best-colleges/lehigh-university-3289/overall-rankings
72. *A history of women at Lehigh: 1921–1971.* (2016). Retrieved December 29, 2016, from www.lehigh.edu/~in40yrs/history/1921-1971.html
73. *Undergraduate Admissions.* (2016). Retrieved December 29, 2016, from http://www1.lehigh.edu/admissions/undergrad
74. *Mission Statement.* (2016). Retrieved October 13, 2017, from http://catalog.lehigh.edu/missionstatement/
75. Johnson, J. (2012, December 22). U-Va. Provost John Simon's defining moment. *The Washington Post.* Retrieved December 29, 2016, from www.washingtonpost.com/local/education/u-va-provost-john-simons-defining-moment/2012/12/22/21faadb2-4b72-11e2-a6a6-aabac85e8036_story.html?utm_term=.80ffa4db4714
76. Ibid.
77. Harbrecht, L. (2016). *For Lehigh, a 'path to prominence.'* Retrieved January 11, 2017, from http://www1.lehigh.edu/news/lehigh-%E2%80%98path-prominence%E2%80%99
78. Simon, J. D. (2016). *What I'm reading: 'A dance to the music of time.'* Retrieved January 11, 2017, from www.chronicle.com/article/what-i-m-reading-a-dance/236553?cid=rc_right
79. *The principles of our equitable community.* (n.d.). Retrieved December 26, 2016, from www.lehigh.edu/~inprv/initiatives/PrinciplesEquity_Sheet_v2_032212.pdf
80. Odi, H. (2016, December 20). Personal communication.
81. *A message from the president regarding Lehigh's inclusive community.* (2016). Retrieved January 11, 2017, from http://www1.lehigh.edu/news/message-president-regarding-lehighs-inclusive-community
82. *Lehigh at a glance.* (2016). Retrieved December 27, 2016, from http://www1.lehigh.edu/about/glance
83. *Diversity.* (n.d.). Retrieved December 27, 2016, from www.amherst.edu/amherst-story/diversity
84. Odi, H. (2016, December 22). Personal communication.
85. Chun, E., and Evans, A. (2015). *The department chair as transformative diversity leader: Building inclusive learning environments in higher education.* Sterling, VA: Stylus Publishing. Please note that in a sample of 98 chairs in four-year, master's level and research universities conducted for this study, only five department chairs or 5 percent were Asian American.
86. Letter dated September 26, 2014, from Vicki Piel, Team Leader/Supervisory Attorney, United States Department of Education, Office for Civil Rights to Kevin Clayton, Interim President, Lehigh University. (2014). Retrieved December 2, 2016, from http://www2.ed.gov/about/offices/list/ocr/docs/investigations/more/03142021-a.pdf.

See also Lin, A. (2014, January 23). Lehigh U. being investigated for 'racially hostile environment.' *DiversityInc*. Retrieved April 7, 2017, from www.diversityinc.com/news/lehigh-u-investigated-racially-hostile-environment/

87. Kingkade, T. (2014, October 2). Lehigh University avoids sanctions in federal investigation of racist incidents. *The Huffington Post*. Retrieved December 28, 2016, from www.huffingtonpost.com/2014/10/02/lehigh-federal-investigation-racism_n_5922080.html
88. Wojcik, S. M. (2013). *Lehigh university alumna files complaint with feds, says discrimination not taken seriously at school.* Retrieved April 24, 2017, from www.lehighvalleylive.com/bethlehem/index.ssf/2013/11/lehigh_university_alumna_files.html
89. Cavanaugh, C. (2013). *UPDATE: graffiti, egging of Lehigh University's UMOJA house sparks calls for action on campus.* Retrieved April 24, 2017, from www.lehighvalleylive.com/thebrownandwhitenews/index.ssf/2013/11/lehigh_universitys_umoja_house.html
90. Simon, J. D. (2015, December 22). *Message to campus regarding results of the Lehigh Survey.* Retrieved December 26, 2016, from http://www1.lehigh.edu/president/speeches/message-president-simon-lehigh-survey-update
91. Clayton, K. L., and Farrell, P. V. (2015, January 26). *Update to campus on the OCR voluntary resolution agreement.* Retrieved December 29, 2016, from www.lehigh.edu/~inis/diversity/1.15%20Update%20To%20Campus%20on%20the%20OCR%20Voluntary%20Resolution%20Agreement.pdf
92. Mallett, K. (2015, January 26). Civil rights investigation prompts university action for inclusion, diversity education. *The Brown and White*. Retrieved December 28, 2016, from http://thebrownandwhite.com/2014/10/24/lehigh-civil-rights-investigation/
93. Shortell, T. (2014). *Former Lehigh University football player sentenced to Northampton county prison.* Retrieved April 24, 2017, from www.lehighvalleylive.com/bethlehem/index.ssf/2014/04/former_lehigh_university_footb_1.html
94. Brown and White Staff. (2013). *Lehigh University student protest calls for administrative action.* Retrieved April 24, 2017, from http://trocaire.ny.safecolleges.com/training/player/AB30B614-937F-11E5-BE26-2EAB26805388/C61D6736-244E-11E7-B5E0-6AB669472AE6/
95. *Memo from president and provost: Update on work to improve diversity and inclusion.* (2013). Lehigh University. Retrieved April 8, 2017, from https://advance.cc.lehigh.edu/news/memo-president-and-provost-update-work-improve-diversity-and-inclusion
96. *The Lehigh survey report.* (2016). Retrieved December 30, 2016, from www.lehigh.edu/~inprv/communications/pubsreports.html#lehighsurvey
97. Ibid.
98. Ibid.
99. Chang, C. (2016, February 10). Separate but unequal in college Greek life. *The Century Foundation*. Retrieved December 29, 2016, from https://tcf.org/content/commentary/separate-but-unequal-in-college-greek-life/. For example at Princeton, one of the few institutions in which the Student Government has collected demographic data on the Greek system, approximately three-quarters of fraternity and sorority members are white compared to 47 percent of the student body. Less than 5 percent of Princeton's Greek life members were from lower- or middle-income families.
100. Chun and Evans. (2016). *Rethinking cultural competence in higher education.* See also Museus, S. D., and Harris, F. (2010). Success among college students of color: How institutional culture matters. In T. E. Dancy, II (Ed.), *Managing diversity: (Re) visioning equity on college campuses* (pp. 25–44). New York: Peter Lang Publishing.
101. Williams, D. A., Berger, J. B., and McClendon, S. A. (2005). *Toward a model of inclusive excellence and change in postsecondary institutions.* Retrieved January 3, 2017, from www.aacu.org/sites/default/files/files/mei/williams_et_al.pdf

102. Ibid.
103. *Advance.* (2017). Retrieved April 24, 2017, from https://advance.cc.lehigh.edu/
104. Chun and Evans. (2016). *Rethinking cultural competence in higher education.*
105. Bronstein, P. (1993). Challenges, rewards, and costs for feminist and ethnic minority scholars. *New Directions for Teaching and Learning*, 53, 1–104. See also Bonner, F. (2016). *The unbearable whiteness of teaching.* Retrieved January 2, 2017, from https://medium.com/voices-on-campus/the-unbearable-whiteness-of-teaching-ef6d8a55a3a3#.ribczo8hd
106. Bonner, F. (2016). *The unbearable whiteness of teaching.*
107. Hochbein, K. (2016). *A rally for inclusion.* Retrieved January 3, 2017, from http://www1.lehigh.edu/news/rally-inclusion
108. *Tackling tough topics together: A brown bag discussion.* (2016). Retrieved January 3, 2017, from http://go.activecalendar.com/lehighu/event/tackling-tough-topics-together-a-brown-bag-discussion/
109. *UMD undergraduate student profile.* (2016). Retrieved November 16, 2016, from https://irpa.umd.edu/CampusCounts/Enrollments/stuprofile_allug.pdf
110. Ibid. See also *Cultural diversity report 2015: University of Maryland, college park narrative.* (n.d.). Retrieved October 8, 2016, from www.irpa.umd.edu/Publications/Reports/cult_div_rpt_2015.pdf/
111. *Cultural diversity report 2015.* (n.d.).
112. Ibid.
113. Orfield, G., Ee, J., Frankenberg, E., and Siegel-Hawley, G. (2016). *Brown at 62: School segregation by race, poverty and state.* Retrieved November 16, 2016, from www.civilrightsproject.ucla.edu/research/k-12-education/integration-and-diversity/brown-at-62-school-segregation-by-race-poverty-and-state
114. *The first 100 years: 1956–1956.* (n.d.). Retrieved November 16, 2016, from http://reslife.umd.edu/programs/programresources/bulletinboard/files/diversity/diversityhistoryumcp.doc
115. *Access is not enough: Introduction.* (n.d.). Retrieved November 21, 2016, from http://umd.edu/commissions/PCEMI/aboutus/reports/Access/introduction.txt
116. Supreme Court Denies Review of Minority Scholarship Program. (1995). *Civil Rights Monitor.* Retrieved October 10, 2016, from www.civilrights.org/monitor/vol8_no1/art2.html
117. The Banneker ruling: Two years later scholarships for blacks are drying up. (1997). *The Journal of Blacks in Higher Education*, 14, 38–40.
118. Folkenflik, D., and Denniston, L. (1995, May 23). Blacks-only aid program dies after justices refuse review college park's Banneker scholarships. *The Baltimore Sun.* Retrieved November 21, 2016, from http://articles.baltimoresun.com/1995-05-23/news/1995143006_1_banneker-black-students-university-of-maryland
119. The Banneker ruling. (1997).
120. De Vise, D. (2009). *U-Md. Students protest official's firing.* Retrieved November 16, 2016, from www.washingtonpost.com/wpdyn/content/article/2009/11/05/AR2009110502997.html
121. Walker, C. (2010, January 1). Budget cuts at college park create unrest: UM students, faculty say they feel left in the dark. *The Baltimore Sun.* Retrieved November 22, 2016, from http://articles.baltimoresun.com/2010-01-01/news/bal-md.collegepark01jan01_1_budget-cuts-cuts-at-college-park-state-budget
122. Cook, B., and Kim, Y. (2012). *The American college president 2012.* Washington, DC: American Council on Education.
123. About the President. (2015). *University of Maryland.* Retrieved October 10, 2016, from www.president.umd.edu/administration/about-president

124. *Transforming Maryland: Expectations for excellence in diversity and inclusion.* (2010). Retrieved October 10, 2016, from www.provost.umd.edu/Documents/Strategic_Plan_for_Diversity.pdf
125. Ibid.
126. Dance, S. (2015, November 19). Towson, Hopkins and U-Md. students join protest movement for racial inclusion. *The Baltimore Sun.* Retrieved October 10, 2016, from http://touch.baltimoresun.com/#section/-1/article/p2p-85096634/. Shapiro, T. R. (2015, December 7). U-Md. board of regents votes to strip Byrd name from football stadium. *The Washington Post.* Retrieved October 8, 2016, from www.washingtonpost.com/news/grade-point/wp/2015/12/07/university-of-maryland-president-recommends-changing-name-of-byrd-stadium-citing-legacy-of-racism/
127. Svrluga, S., and Wiggins, O. (2015, December 7). University of Maryland president recommends changing name of Byrd stadium, citing legacy of segregation. *The Washington Post.* Retrieved October 8, 2016, from www.washingtonpost.com/news/grade-point/wp/2015/12/07/university-of-maryland-president-recommends-changing-name-of-byrd-stadium-citing-legacy-of-racism/
128. Ibid. See also Svrluga, S., Anderson, N., and Hedgpeth, D. (2015, March 13). A U-Md. fraternity brother's e-mail full of racial slurs has something to offend just about everyone. *The Washington Post.* Retrieved October 8, 2016, from www.washingtonpost.com/news/grade-point/wp/2015/03/13/a-u-md-fraternity-brothers-e-mail-full-of-racial-slurs-has-something-to-offend-just-about-everyone/
129. Meichensehrdbk@gmail.com. (2015, March 27). UMD students march in protest on fraternity row in light of racist, sexist email: About 100 gather for stamp student union sit-in and share list of demands. *The Diamondback.* Retrieved October 10, 2016, from www.dbknews.com/archives/article_11c2ae14-d404-11e4-9a95-fb69e12eea25.html
130. Svrluga, S., and Anderson, N. (2015, April 2). Many at U-Md. upset after its president says slur-filled e-mail didn't violate school policy. *The Washington Post.* Retrieved October 8, 2016, from www.washingtonpost.com/news/grade-point/wp/2015/04/02/many-at-u-md-upset-after-its-president-says-slur-filled-e-mail-didnt-violate-school-policy/
131. Ibid.
132. Kohli, S. (2015). *Read this university president's candid reaction on Twitter to frat-house racism on his campus.* Retrieved October 8, 2016, from http://qz.com/#362466/read-this-university-presidents-candid-reaction-on-twitter-to-frat-house-racism-on-his-campus/
133. Gray, E. (2015, April 2). University of Maryland won't expel student who sent racist, sexist email. *Time Magazine.* Retrieved October 7, 2016, from http://time.com/3769602/maryland-university-student-racist-sexist-email/#
134. *University of Maryland—College Park.* (n.d.). Retrieved October 8, 2016, from https://colleges.niche.com/university-of-maryland——college-park/
135. *Cultural diversity report 2015.* (n.d.).
136. Ibid.
137. Szrom, M. (2016). The importance of identity: How the University of Maryland uses self-discovery to explore diversity. *Insight Into Diversity.* Retrieved October 8, 2016, from www.insightintodiversity.com/the-importance-of-identity-how-the-university-of-maryland-uses-self-discovery-to-explore-diversity/
138. Ibid.
139. Escobar, K. (2016). *College Park and UMD officials address black lives matter at Maryland discourse event.* Retrieved October 8, 2016, from www.dbknews.com/2016/02/26/black-lives-matter-umd-maryland-discourse/

5
FUTURE ASPIRATIONS AND EXPECTATIONS

> But what if we thought of change not just as a chain of events? Instead what if we thought of change also as a process that, like the ocean itself never stops moving?
>
> Jeff Chang, 2014, p. 6[1]

As we conclude the study, the question remains as to whether real progress has been made in leading a diversity culture shift within institutions of higher education. Is cultural change occurring and proceeding at a significant pace? We have traced the promising trajectory of a number of institutions that actively are promoting diversity organizational learning through multiple, interconnected initiatives. Few institutions, however, have reached the stage of institutionalizing diversity learning programs across the varied contours of the campus ecosystem. Some colleges and universities are still in early phases of experimentation and are building buy-in among the larger schools on their campuses, while others have created unit-level diversity planning models that roll-up to a centralized strategic diversity plan. A number of Ivy League institutions have devoted significant new funding resources to diversity initiatives. But progress on a diversity agenda still can be an uphill battle when faced with regressive political forces and a divisive national climate. Since institutions of higher education reflect and replicate the dominant culture in society, cultural change is a gradual process influenced by shifting demographic tides and the addition of new cultural elements.[2]

Student demonstrations have been a major force in the process of diversity transformation and especially in efforts to enhance diversity education. The courageous voices of students protesting exclusionary campus climates and pressuring for change have called attention to the need for more democratic and inclusive learning environments. The case studies offer remarkable examples of student activism that have had a direct impact on university administration in terms of concrete programmatic and leadership changes. Yet we have also seen that as students graduate and coalitions dissolve, the long-term impact of student efforts will be difficult to sustain.

Recall the "law of social inertia" based on Newtonian principles that requires a countervailing force to bring about sustained change in institutionalized systems.[3] Absent such a force, the status quo and organizational inertia with respect to diversity change will prevail. We have argued throughout this book that committed diversity leadership and ongoing support by the president or chancellor as well as the Board of Trustees are essential for sustained cultural change. At the University of Tennessee at Knoxville (UTK) and the University of Missouri at Columbia ("Mizzou"), tenured faculty leaders and academic senates have joined student voices to represent a significant countervailing force. Nonetheless, as several diversity officers have pointed out, the hope is that institutions will continue to accelerate the change process without the impetus of student pressures. As they have stated, the goal is to be proactive, not reactive. By contrast, when diversity learning programs are sidelined as a peripheral part of the university or college agenda, these programs can dissolve into symbolic activities that may be redundant, fragmented, and siloed.

A number of significant obstacles can short circuit the progress of diversity learning. Colleges and universities are essentially conservative institutions and late adapters to changing social conditions.[4] College presidents themselves, particularly those in public institutions, may be subject to powerful pressures from conservative legislatures that exercise some degree of control through budgetary allocations, as we have seen in the UTK case study. Alumni support can present conservative perspectives on student demonstrations. Student social networks such as the Greek system or even eating clubs can make it difficult to achieve greater integration in the campus community. As the final arbiter of university decision-making, we have also seen how Boards of Trustees can override institutional perspectives on free speech, academic due process, or the naming of buildings. The frequent turnover of top institutional leaders can cause the mandate for diversity change to erode and alter the priorities of key administrative departments. Within politicized campus environments, internal cultural opposition to diversity sometimes serves as a potent and clandestine brake on progress.

Our findings also indicate that while institutions of higher education frequently have issued statements about the importance of diversity, these statements may not have trickled down to a discernible level in the day-to-day experiences of diverse administrators, staff, and students. In such cases, diversity can be relegated to an overall marketing and branding effort designed to showcase the institution in a positive light, rather than accorded priority in the institution's strategic agenda. When a gap between rhetoric and reality occurs, skepticism and fear of reprisal arises, as evidenced by a number of individuals we interviewed who did not wish to be identified as well as by the reluctance of chief diversity officers to participate in this study.

Given the relative newness of the Chief Diversity Officer position, the expectations, responsibilities, and scope of these positions are still subject to considerable variation and lack standardization. While the National Association of Diversity Officers (NADOHE) has set forth expectations and standards for CDOs, the requirements articulated in these standards indicate a foundational level of knowledge and skills. Described as a "formative advancement" leading toward the "increased professionalization" of the CDO role, all 12 NADOHE standards focus on understanding, awareness, and communication rather than the ability to lead cultural change, deploy effective organizational learning, collaborate with key stakeholders, and take action to ensure accountability for diversity progress.[5] Since, as we have indicated at the outset, diversity is not a quick fix, the next iteration of these standards likely will accent the action-oriented expectations of the CDO role.

Almost all the diversity officers interviewed for this study are untenured and those who are tenured were previously full-time faculty members. Without employment protection and when serving in "at will" positions, diversity officers may lack the clout or organizational support to instigate significant cultural change. We have noted earlier the reluctance of diversity officers to participate in our survey and this reluctance extended in some instances to other members of the administration. Recall how Yekim, an Afro-Latino CDO in a western university, put it, "The role is in itself a risk. . . . the reality is that this work challenges power." During the course of this study, we have noted turnover of several chief diversity officers. To complicate the picture, diversity officers often represent the lone minority presence in a largely white administrative hierarchy and this isolation often suggests a symbolic role without the authority to challenge existing power-based realities. Since a key attribute needed for this work is expertise in change management, diversity officers have to be equipped to keep the cultural change process moving and flowing toward the goal of institutional transformation.

Many of the diversity officers in our study hold Ph.D.s, an almost threshold requirement for gaining acceptance among faculty. In our view, the creation of faculty diversity officer positions held by individuals with tenure heightens the potential to affect cultural change at the academic core. At the same time, a number of diversity officers we interviewed do not have even one fully dedicated staff member devoted to training and work with only a small contingent of administrative and professional support staff. As a result, these leaders are limited in the ability to provide comprehensive organizational learning programs and often spend part of their own time conducting seminars or turn to outside consultants to conduct workshops.

One of the most salient strategies for diversity learning is the development of collaborative efforts among institutional offices such as Student Life, Human Resources, and Faculty Affairs. Nonetheless, the expertise of Human Resources (HR) in change management often has been overlooked and underutilized. HR has the capacity to strengthen professional development programs that include both faculty and staff, work with management and supervisors in creating diverse departmental cultures, and redesign institutional processes to enhance diversity organizational capability. Student Affairs is also an instrumental player in articulating student concerns, creating programs, centers, and safe spaces that foster inclusion, and providing developmental co-curricular opportunities. Faculty Teaching and Learning Centers offer the potential to move beyond instructional development to organization development in alignment with institutional goals for diversity and inclusion. In sum, developing collaborative synergies across often territorially drawn institutional lines will help consolidate, coordinate, and deploy intentional programs for diversity learning for administrators, faculty, staff, and students.

New pluralistic leadership models are needed to reshape and replace hierarchical and elite modes of governance and build inclusive environments for diverse faculty, staff, administrators, and students. As Feagin and Ducey point out, asymmetrical power structures reproduce social systems in which "more powerful people repeatedly and profitably impose their interests and goals on less powerful people."[6] The continued dominance of white, male, heterosexual perspectives in university and college administration has failed to foster a representative bureaucracy that is responsive to the needs of diverse students. New leadership models that build teamwork, value cultural differences, create trust-based rather than fear-based environments, and reflect diverse constituencies are foundational components needed for substantive organizational change to occur.

Based on the research, survey findings, and case studies in this book, we have identified 20 key features of successful diversity organizational learning programs as follows:

1. Committed and academically focused leadership direction from the president or chancellor;
2. Sustained strategic direction and in-depth monitoring of diversity outcomes by the board of trustees;
3. The development of a common vocabulary for defining diversity, inclusion, organizational learning, and related concepts such as cultural competence;

4. Prioritization of diversity organizational learning in alignment with institutional mission and goals and articulated in strategic planning documents;
5. A research-based approach to diversity learning based on a change model such as the framework of Inclusive Excellence;
6. A focus on the relation of diversity education to student learning outcomes;
7. Structural coordination and consolidation of diversity learning programs;
8. Use of data-driven models to assess diversity progress;
9. Development of diversity education policies and accountability measures;
10. Allocation of recurring funding resources for diversity organizational learning and appointment of dedicated training specialists in diversity offices to conduct education efforts;
11. Active engagement of campus governance structures and constituencies including faculty senates;
12. Critical expertise among diversity leaders in change management to address the ways in which sociohistorical forms of exclusion are reproduced in the educational context;
13. A focus on institutional processes such as hiring, promotion, compensation, and termination in terms of equity and the ways in which micro-aggressions, stereotypes, and implicit bias can affect outcomes;
14. Willingness to engage in candid self-evaluation processes on diversity progress with input from stakeholders using climate studies and other empirical measures;
15. Mapping of current diversity learning initiatives in terms of content and objectives with a coordinated approach across all divisions and units;
16. An iterative, research-based diversity education plan with gap assessment, benchmarking, measurable outcomes, longitudinal evaluation of transferability of learning, and designated accountability within defined timeframes;
17. Learning frameworks that bridge disciplinary divides such as through interdisciplinary programs;

18. Rewards and incentives that recognize substantial contributions toward diversity learning and practices at the individual, team, unit, college/school, and institutional levels;
19. Recognition of diversity-related leadership and scholarly contributions in promotion and tenure guidelines;
20. Institutionalization of coordinated programs of diversity learning across decentralized organizational units that include diverse constituencies of faculty, administrators, staff, and students.

The research shared in this book offers considerable hope for building a positive and progressive pathway toward diversity organizational learning. There is no time to wait. Student activists representing the new Joshua generation have galvanized the process of diversity change. There is still some distance to travel to ensure that campuses close the gap between institutional mission and day-to-day experiences of diversity though inclusive practices, processes, and interactions. It is now up to colleges and universities to sustain and even accelerate the momentum that student demonstrations have generated. From this perspective, diversity and inclusion are not tangential considerations, but are fundamental to the success of institutions of higher education. As Edmund O. Wilson writes, "To move forward is to concoct new patterns of thoughts, which in turn dictate the design of the models. . . . Easy to say, difficult to achieve."[7] A diversity culture shift means setting new norms for interactions, behaviors, and learning outcomes.

Leading a diversity culture shift requires strategic courage and navigational skill to identify strategies to surmount covert and overt waves of opposition. It is a continuous, evolutionary process of orchestrating institutional structures, processes, and practices with the goal of creating a vibrant and inclusive learning, living, and working community that values, nurtures and respects the contributions of all its members. Consider, in conclusion, the words of Barack Obama in his victory speech upon reelection in 2012:

> Our university, our culture are all the envy of the world, but that's not what keeps the world coming to our shores. What makes America exceptional are the bonds that hold together the most

diverse nation on Earth, the belief that our destiny is shared—that this country only works when we accept certain obligations to one another and to future generations.[8]

Notes

1. Chang, J. (2014). *Who we be: A cultural history of race in post-civil rights America*. New York: St. Martin's Press.
2. Feagin, J. R. (2016). *How blacks built America: Labor, culture, freedom, and democracy*. New York: Routledge.
3. Feagin, J. R. (2006). *Systemic racism: A theory of oppression*. New York: Routledge, p. 34.
4. See Chun, E., and Evans, A. (2015). *The department chair as transformative diversity leader: Building inclusive learning environments in higher education*. Sterling, VA: Stylus Publishing.
5. Worthington, R.L., Stanley, C.A., and Lewis, W.T. (2014). *Standards of professional practice for chief diversity officers*. Retrieved October 24, 2017 from www.nadohe.org/standards-of-professional-practice-for-chief-diversity-officers
6. Feagin, J. R., and Ducey, K. (2017). *Elite white men ruling: Who, what, when, where, and how*. New York: Routledge, p. 19.
7. Wilson, E. O. (1999). *The diversity of life*. New York: W. W. Norton.
8. Obama, B. (2012, November 7). *Audio and transcript: Obama's victory speech*. Retrieved May 4, 2017, from www.npr.org/2012/11/06/164540079/transcript-president-obamas-victory-speech

APPENDIX A

DIVERSITY OFFICER SAMPLE

Although a number of interviews were conducted with institutional leaders, faculty, administrators, and students, the information provided in this appendix pertains to individuals with the specific designation of diversity officer. The sample included chief diversity officers (CDOs) as well as faculty diversity officers and officers located in large schools and colleges. Of the 37 diversity officers in the survey sample, 32 individuals indicated a willingness to be interviewed. The interviews were conducted during the time period from June 2016 to April 2017. Most interviews were conducted over the telephone and typically lasted one hour, although several were conducted in person. Before each interview, informed consent was obtained to record the interview for note-taking purposes and the participants were provided transcribed passages selected for use in the study for their review and consent.

The demographic background of the survey participants included 21 females and 16 males, with 31 African Americans, 3 Asian or Asian Americans, and 3 whites with 4 individuals identifying as Hispanic in ethnicity.

The survey sample included institutions in all geographic areas with 17 in the East, 11 in the Midwest, 2 in the South, 2 in the Southwest, and 5 in the West. The types of institutions represented in the survey sample include 16 in public research universities, 12 in private research universities, 3 in private master's institutions, 1 in a public master's institution, and 5 in private four-year colleges.

APPENDIX B

DIVERSITY ORGANIZATIONAL LEARNING ASSESSMENT MATRIX

Directions: This matrix will assist institutional diversity leaders in assessing progress in developing systematic organizational learning programs. Key areas such as leadership commitment, resources, staffing, and organizational design will impact the ability to move a diversity cultural shift forward.

Rating scale: Please rate the approximate phase you believe your institution has reached in each of these areas as follows:

1 = Mobilization 2 = Implementation 3 = Institutionalization

Area	*Factors*	*Assessment*	*Comments*
Leadership commitment	1. Addresses diversity organizational learning in strategic planning documents		
	2. Regularly assesses processes and programs for equity, diversity, and inclusion		
	3. Establishes accountability for diversity progress through budgetary or annual review mechanisms		
	4. Measures Cabinet officers on key performance indicators of diversity progress		
	5. Conducts regular climate study or diversity audit		

Area	*Factors*	*Assessment*	*Comments*
Organizational design	1. Promotes collaboration on diversity learning across organizational units		
	2. Minimizes diversity silos through oversight and coordination of diversity efforts		
	3. Delegates authority to diversity-related positions		
	4. Creates structures for stakeholder input		
Resources	1. Provides specialized positions for diversity professional development within diversity office(s)		
	2. Provides a regularized budget for diversity professional development		
	3. Allocates sufficient financial resources for institutional diversity learning and education programs		
Inclusive Excellence	1. Addresses the need for enhanced organizational learning as part of the IE framework		
	2. Invokes consideration of how systems of power have defined excellence		
	3. Moves beyond conversations and dialogue to address structural barriers to diversity and inclusion		
Diversity culture	1. Promotes inclusive behaviors and interactions on a day-to-day basis within the university/college community		
	2. Includes diverse faculty, administrators, and staff in decision-making		
	3. Ensures equitable resource distribution		
	4. Offers formal and informal networks of support for diverse faculty, administrators, and staff		

Area	*Factors*	*Assessment*	*Comments*
Incorporation of recognition and rewards for diversity learning	1. Includes diversity as a criterion in evaluation processes for faculty, administrators, and staff		
	2. Incorporates diversity-related criteria in faculty promotion and tenure processes		
	3. Provides recognition to units that have made substantive diversity progress		
	4. Supports and encourages participation in diversity education programs		
	5. Promotes diversity research agendas and scholarship		
Overall Assessment			

Synopsis

What are the principal barriers to diversity organizational learning (e.g. staffing, resources; historical legacy; leadership commitment, cultural resistance, etc.)?

____

____

What specific strategies would help your institution address these barriers?

____

____

INDEX

Page numbers in bold indicate tables.

Made in the USA
Columbia, SC
29 July 2020

14970340R00143